Case
Dismissed!

Taking Your Harassment
Prevention Training to Trial

Carol M. Merchasin, Esq.
Mindy H. Chapman, Esq.
Jeff Polisky, Esq.

Defending Liberty
Pursuing Justice

Section of State and Local Government Law
Tort Trial and Insurance Practice Section
American Bar Association

Cover design by Cathy Zaccarine.

07 06 05 04 03   5 4 3 2 1

Library of Congress Cataloging-in-Publication Data

Merchasin, Carol, 1946–
    Case dismissed : taking your harassment prevention training to trial /
Carol Merchasin, Mindy Chapman, Jeff Polisky.
      p. cm.
Includes index.
   ISBN 1-59031-156-6
   1. Sexual harassment—Law and legislation—United States.
2. Harassment—United States—Prevention. I. Chapman, Mindy.
II. Polisky, Jeff. III. Title.

KF4758.M47 2003
344.7301'133—dc21

                                      2003000378
                                          CIP

Discounts are available for books ordered in bulk. Special consideration is given to state bars, CLE programs, and other bar-related organizations. Inquire at ABA Publishing, Book Publishing, American Bar Association, 750 North Lake Shore Drive, Chicago, Illinois 60611.

www.abanet.org/abapubs

To the thousands of participants
we have trained over the years.
They have taught us well.

Acknowledgments

We have watched endless television award shows, wondering all the while how people who are otherwise articulate and talented can babble on and on with a never-ending list of people to whom they are deeply indebted. We must admit that now, having produced our own little work of art, we understand the urge and the need. We hope you will indulge us as we express our gratitude. If you simply are too bored, just move on to Chapter 1 without guilt. We forgive you. We would probably do the same.

We are indebted to our partners and colleagues at Seyfarth Shaw for their support and encouragement. We wish to especially thank James Baird who suggested that we team up with the American Bar Association to write this book; Laura Lindner who lost many billable hours critiquing our drafts; David Bowman and Martha Michael Gates for "holding down the fort" while we were writing; and members of the Alternative Resources Group for their research.

It would not have been possible to turn hundreds of cocktail napkins and Post-its into coherent text without the patience and dedication of members of the Seyfarth Shaw At Work team, including Rick Ariano, Lauren Christoff, Nancy Dellamore, Valerie Molloy Fox, Tina Simmons, and Dawn Zerbs.

We also wish to thank the hundreds of trainers who traveled far and wide to help us bring our training to our clients every day, including Philippe Weiss who read drafts in taxicabs, faxed suggestions from hotels, and phoned us with his comments while waiting for his luggage at airports. We also must express our appreciation to the staff at the American Bar Association, including Rick G. Paszkiet, executive editor, Catherine Flanagan, marketing director, and Maureen Grey, book production manager.

Carol wants to express her love and gratitude to: Her husband, Robert, for his patience with her endless requests to either 1) leave her alone or 2) read the manuscript once again. His gentle grammatical corrections are noted (although we still cannot believe it is true that a sentence cannot end in a preposition or there is such a thing as a run on or repetitive, really repetitive sentence); to her children Sean and Jessica for providing her with the greatest training ground ever for being a trainer and for helping her to have a sense of humor; to her mother, Gloria, for training her so well; to her father, Stan, she knows he is looking down at her beaming with pride. To her favorite dogs, Max and Penny, who are a constant source of inspiration on how easy it would be to sleep throughout the day.

Mindy expresses gratitude for her blessings . . . *and love to each of them:* To Richard, her husband, for his extraordinary friendship, strength and encouragement (not to mention his ability to simultaneously practice law, read the drafts and cook dinners—well, at least "order in"); to her beloved children, Emily and Danny, for being the greatest teachers of love, kindness, compassion, and fun; to her devoted mother, Marcia Tarre, for supporting her courtroom dreams as a child and making them come true as an adult; to her father, Marshall Tarre, her cherished angel above, and for a very supportive family: Joseph, Renee, Kenneth and Lauren Hecht; Louise, James, Toosheyah, Robert, Sandy, Alex and Kayla Chapman; Judith Elting and family.

Stop playing the music we are almost done. . . .

Jeff wishes to thank his family, the Academy, his agent, and everyone he stepped on as he climbed the ladder. Oh, and his invisible friend Rollo.

Contents

About the Authors

Carol Merchasin, Esq., is the national director of training for Seyfarth Shaw at Work®, a subsidiary of Seyfarth Shaw, one of the nation's largest labor and employment firms. Although she once practiced employment law defending management, Carol has made the business of providing high-energy, interactive legal compliance training a more than full-time occupation. Carol speaks extensively on training and legal compliance issues and is on the faculty of the Massachusetts Commission Against Discrimination Train-the-Trainer program.

Mindy Chapman, Esq., is the executive vice president of training and development for Seyfarth Shaw at Work® and a lawyer who has represented plaintiffs in many areas of the law, including employment cases. Mindy has an extensive background in training and is a master trainer, blending the law and humor to facilitate learning. Mindy believes you can take any legal topic, identify the main teaching points, and deliver them in an interactive way so that participants can acquire new skills—hence, her additional hat as a course developer. Mindy is the Project Manager for many clients' legal training initiatives, including extensive nationwide compliance programs that she manages in accordance with Equal Employment Opportunity Commission (EEOC) conciliation agreements. Together, Carol and Mindy have developed innovative, high-energy training that is being delivered across the country for clients large and small. A number of their training programs have been approved by the EEOC for use under conciliation agreements.

Jeff Polisky, Esq., is Seyfarth Shaw at Work®'s lawyer turned creative director and the producer of Seyfarth Shaw at Work®'s videos and web-based training projects. Jeff is also an accomplished legal researcher, able to find legal cases, statutes, and regulations faster than you can say, "Can you please . . . ?"

Introduction

Does your organization have an extra $20-plus million to give away today? We didn't think so. And even if it did, would you want to give the money away to a noncharitable cause? We didn't think so, either. But it's not just about the money. Can your organization afford to lose the goodwill and name recognition it has worked so hard to achieve? No again? That is probably why you have started reading this book about harassment prevention training. Good first step.

With American businesses spending between sixty and one hundred billion dollars each year on training, workforce or employer-sponsored training has become an industry unto itself. From the boardroom to the plant floor, from supervisors and managers to the rank and file, training programs have permeated the 21st-century working environment. And while the subjects addressed in training programs are a vast and varied lot, training a workforce on employment laws is becoming a key staple in any organization's training program.

In response to sexual harassment's notoriety, partially the result of a number of highly publicized sexual harassment verdict awards and settlements (with six and seven figures after the dollar sign), the sine qua non of employment training is sexual harassment prevention training.[1] In fact, over 87% of all employers train in this area.[2] Employers have begun conducting harassment training to such a degree that it has become part not only of the legal vernacular, but of the workplace vernacular as well. It has invaded the corporate world and established a foothold in the employment culture.

Speaking of the vernacular, the phrases "harassment training" and "harassment policy" are most commonly used. However, "harassment prevention training" and "harassment prevention policy" are undoubtedly more accurate. We use them interchangeably. If you don't know the difference, it's a good thing you're reading this book!

Yet with the recent emergence of harassment training, the inevitable questions arise: What should the content of the training program be? How can training protect an organization from liability or from future harassment complaints? How can I make the program interesting for the audience? How does an organization properly conduct harassment training in a cost-effective way to achieve ample benefits?

From the design of the training course to the most effective class size, it is the intention of this book to take the reader through the nuts and bolts of harassment training and the decisions that need to take place behind the scenes. Our goal is to provide guidance to the Human Resources manager or executive or attorney who is suddenly faced with the challenge of either building a harassment training program from nothing (generally with a cry for help on Friday afternoon and a delivery deadline of Monday morning) or auditing an existing harassment training program to ensure that the program is reaping all the benefits an exemplary training program should provide.

Do also remember that there are other important legal compliance topics that merit training, including modules for supervisors and managers on legally effective interviewing, making nondiscriminatory hiring and promotional decisions, working with the Americans with Disabilities Act and the Family and Medical Leave Act, nondiscriminatory performance management, and—the bane of most HR and legal professionals' existence—performance documentation.

But these are topics for another day.

For today, our plan is to help you create the best harassment training program for your organization. One that will help you create a culture of respect. One that will be comprehensive enough to satisfy legal compliance issues, yet be fun and practical. One that will shield you from liability and keep you from ending up in litigation. But, if you do end up in court, it is there that our goal gets reduced to the bottom line. We want you to hear only two words based on your harassment prevention policy and training:

"Case dismissed!"

Notes

1. While we are saying *sexual* harassment training, the reality is that harassment training should focus on all forms of harassment protected by federal law, including harassment related to race, religion, national origin, disability, age, gender, and any other characteristics such as sexual orientation that are protected by various state laws. Sexual harassment is merely the most "famous" of the possibilities. While discussion of sexual harassment issues often is the emphasis of a training program (and sometimes is the reason that a training initiative is undertaken), we stress that the employer's policy, and the training, should cover harassment based on other protected characteristics as well. As we will show you in this book, this is not a difficult undertaking.

2. The annual 2002 Industry Report published by *Training* magazine showed that 87 percent of companies offered sexual harassment training programs.

How to Use This Book

Sometimes, in a very busy, pressure-filled world, with personal and professional demands—telephone calls, pagers, faxes, cell phones, e-mails and letters, cars that need gas, groceries that need to be put in carts, laundry that needs to be washed, and dogs that need to be walked—you just have to sit down and write a book. That's what happened to us.

This book was born out of our search for ways to make our own harassment prevention training better—and our frustration at not finding them. We searched the bookstores, the Internet, and even professional magazines. But we found very little that gave us the content in an easily digestible format without overwhelming legal jargon. And there was even less about tools and exercises we could use to deliver harassment training in ways that would be not just legally sound but fun and interactive.

Harassment training is not like other training. When you train on technical skills, or team-building, you do not have to worry about the legal implications of what you are doing, or that you may be called upon to defend your curriculum in a court of law. In addition, the issues of harassment can come with some heavily loaded emotional baggage. We are, for example, frequently called upon to do harassment training while a lawsuit or complaint is pending and everyone is entrenched in various positions. Even when a claim is not pending, there are those who feel that the world has changed and not for the better. All of this has led us to the conclusion that training on this topic is complex. On the one hand, we must understand the somber legal issues and the heavy liability concerns we are exposed to. On the other hand, we must be upbeat, highly energetic, and very interactive to successfully engage participants so they can understand concepts and acquire new skills.

Not having found the useful, practical book that we longed for, we decided to write our own. Why us? For one thing, we are employment lawyers, so we know the content well. In addition to being lawyers, we, along with our cadre of talented and dedicated trainers, have had the good fortune to have trained more than 200,000 workers on harassment prevention in the last two years. What that means is that we have collected a number of useful tricks, practical tips, and tools (and we have seen the inside of just about every major airport in America). Like the magpies that most trainers are, we have used those tips, tricks, and tools, sometimes successfully, sometimes not. From executive suites to frontlines, we have tried one thing and then another until, like Goldilocks, we hit on what felt "just right." Feel free to benefit from our mistakes.

That does not mean, of course, that everything that works for us will work for you. Because if there is one thing that we know for sure, it is that trainers train best when they are operating within their own comfort level. We train all the time, and sometimes together. But there are tools that one of us may use and the other one doesn't, and vice versa. Pick those tricks, tips, and tools that you want to try on and see if they suit you. If they fit and work for you—great! If a tool or trick doesn't fit right, take it off and hang it up. On another day, you may feel more adventurous and want to try it again in a different way or try something new. When you do, come back to this book and look again.

Who Should Read This Book?

Now a word about who should use this book.

If you are:

■ An employment lawyer, but have never done training;
■ A trainer who doesn't know the harassment content; or
■ In Human Resources and don't have time to "reinvent the wheel"—

then this book is for you.

If you have plans to:

■ Prepare a Request For Proposal (RFP) to outsource harassment training;
■ Review your organization's harassment training; or better yet,
■ Design your organization's harassment training; or best of all,
■ Go out in two weeks to deliver harassment training (and you don't have a clue what you are going to do)—

then this book is for you. Simply put, if you have any responsibility to purchase, monitor, review, or deliver harassment training, this book will quickly help you get it all together.

Roadmap to This Book

We have tried to structure this book in the same logical pattern that most people will follow to tackle the task of creating and conducting harassment prevention training. In Chapter 1, we put on our lawyers' hats to answer the question, Why do employers need harassment prevention training? As an employer's harassment prevention policy is the heart and soul of any training, Chapter 2 discusses the ingredients necessary for a legally sound harassment prevention policy. We make the small leap from the content of the policy to the content of the training program in Chapter 3. Chapters 4 and 5 address how to design a legally sound training course, while still making the program an interactive and memorable learning experience. Chapter 6 focuses on imparting tips to trainers on delivery skills. And, as we started with a legal chapter, so too do we end, as Chapter 7 talks about legal issues that may arise from the training itself, and how to avoid such problems. The appendixes are filled with reference and resource goodies.

Keep on the lookout throughout this book for the following:

 LESSONS LEARNED (at the beginning and end of each chapter)

CHECKLISTS

 **TIPS**

 *TEACHING POINTS*

These are scattered throughout the chapters, providing useful teaching tools and pointing out pitfalls to avoid.

We do have a few disclaimers. What would a book written by lawyers be if it did not have disclaimers?

- First, we are not professionally trained, diplomas-on-the-walls instructional designers. In fact, upon close inspection you will see that this book has no graphs or pie charts. We wish we had at least one for you to skip over. Rather, our expertise in training and developing courses comes mostly from the School of Experience, with some professional training along the way. We have found that the School of Experience has the toughest teachers. And to that end, we have placed ourselves as students to learn from our participants. We pilot all the courses we design and train. We then conduct extensive (and sometimes painful) debriefings with our participants. This allows us to go back to the drawing board and pinch and tweak a course until it gets the highest ratings for content, interaction, and effectiveness. In the end, we realize that putting a course together is like putting on an interactive play. The more dress rehearsals you have, the better the show will be.

■ Second, this book does not purport to give legal advice. If you think you need legal advice, you should definitely contact your employment lawyer.

■ Third, where we have talked about the law in general, and specifically where we have talked about the *Faragher/Ellerth/ Kolstad* defenses (Chapter 2), we have simplified it for readers who are not employment lawyers. Our goal is to give a general overview with enough information to make it clear why harassment training is critical and why harassment training may be a factor at trial. This is not the book you want if you are litigating a harassment claim.

■ Fourth, as you will soon see, there are a number of factors that influence whether an employer will be subject to liability for harassment—from conducting proper investigations to drafting effective policies. While we may briefly discuss a number of these factors, the focus of this book is on training and training alone.

■ Finally, federal, state, and local laws may change over time. Different courts with different judges and different juries interpret the changing laws differently every day. It is hard to wrap your hands around something that is moving. However, the basic laws preventing discrimination and harassment were passed in the last century. And the trend we see is to add more of them. We see no movement going backward to undo them. Therefore, based on that premise, this book is general enough to remain a classic tool in your tool chest for a long, long time.

So, in your busy world of professional and personal demands (telephone calls, pagers, faxes, cell phones, e-mails, letters, cars, groceries, laundry, dogs, etc.), we hope that this book takes away the pressures of harassment training and brings you even a small measure of the satisfaction that training has brought us.

1

Who Wants to Be a Millionaire? The Legal Implications of Harassment Prevention Training

Dear Trainers:

Why do I need to know about the legal issues surrounding harassment prevention training? Isn't it enough to tell our employees to "do the right thing," or "just use common sense"? If everyone acts civilized then no one can sue and become a millionaire off these claims, right? The last thing I have time for is to try to understand a bunch of legal cases. What gives?

Signed,

Who Wants to Be a Millionaire?

. . . .

Dear Millionaire:

Look at the recent Supreme Court cases as a gift! The Court has given your organization the opportunity for a great defense to eliminate liability, or at the very least punitive damages, in claims of harassment. But telling your employees to "do the right thing" or "just use common sense" may not be enough to get you the gift. Knowing more about the *law* of harassment will help you and others in your organization (like your boss) to understand that training is a good investment. It will also help you to develop the best training. Hang in there with us—it won't be too painful, we promise!

Signed,

The Trainers

Overview

It used to be that employers who conducted harassment prevention training had a couple of objectives:

- To avoid litigation by preventing harassment from taking place; and
- To make the workplace a better place.

These goals were great goals, and they still are. But they were limited. For example, you could still be sued. You could still have to go to trial. You could still have liability and punitive damages. Your good motives might not get recognized. Now, however, since 1998, we can add two more very important goals that have been given to us by the United States Supreme Court:

- To be able to raise an affirmative defense to defeat liability in supervisory harassment cases; and
- To minimize the likelihood of being assessed punitive damages.[1]

This means that you need to have a basic understanding of the evolution of harassment law so that you can properly shape the content and structure of your harassment training program. If you have that basic understanding, you will be in the best position possible to meet all your training goals.

What You Need to Know . . .

A Short History of Title VII and Harassment Claims

How did we get to where we are today? Driven by the explosion of employment litigation, the U.S. Supreme Court issued a number of decisions in the last several years, emphasizing the need for employers to take preventive steps in order to avoid Title VII liability and punitive damages. Preventive steps favored by the Supreme Court and other courts include an effective workplace harassment policy, reporting mechanisms, and enforcement and training.[2] With this new emphasis, courts have altered the legal landscape. The logical end result is that an employer's workforce must be well aware of and abide by the organization's policies and procedures. This, then, has become one of the key objectives of harassment training—to raise awareness of an organization's policy and procedures.

What Is Title VII?

The Civil Rights Act of 1964 includes Title VII, which prohibits employment discrimination and states in part:

> It shall be an unlawful employment practice for an employer—to fail or refuse to hire or to discharge any individual, or otherwise to discriminate against any individual with respect to his compensation, terms, conditions, or privileges of employment, because of such individual's race, color, religion, sex, or national origin . . .[3]

While explicitly prohibiting sex discrimination, there is no mention of the term sexual harassment in Title VII or in Title VII's legislative history. But in 1980, the Equal Employment Opportunity Commission (EEOC), the agency that enforces Title VII, issued guidelines interpreting Title VII to forbid sexual harassment as a form of gender discrimination. According to the EEOC:

> Unwelcome sexual advances, requests for sexual favors, and other verbal or physical conduct of a sexual nature constitutes sexual harassment when submission to or rejection of this conduct explicitly or implicitly affects an individual's employment, unreasonably interferes with an individual's work performance or creates an intimidating, hostile or offensive work environment.[4]

The EEOC broke down this definition to conclude that a violation of Title VII may be predicated on one of two types of sexual harassment:

1. Quid Pro Quo Harassment (harassment that involves the conditioning of employment benefits on sexual favors),[5] and
2. Hostile Environment Harassment (harassment that, while not affecting economic benefits, creates a hostile or offensive working environment).[6] This is also commonly referred to as a Hostile Work Environment.

It would take six more years until the highest court of the land first addressed the concept of sexual harassment in a fiery case involving a bank teller and her boss.

The Supreme Court Chimes In

Who knew that when Sidney Taylor at Meritor Savings Bank hired Mechelle Vinson over a quarter of a century ago, the world of employment law was about to change. While Title VII had been passed in

1964 and the term "sexual harassment" already had been recognized by the EEOC and lower courts, it was Taylor's hiring of Vinson for a teller-trainee position at the bank in 1974 that forged the sexual harassment path that we continue to travel on today. For when Vinson was later discharged for excessive use of sick leave, she filed a lawsuit against the bank and Taylor, alleging that she had "constantly been subjected to sexual harassment" by Taylor in violation of Title VII.

On June 19, 1986, in the case of *Meritor Savings Bank v. Vinson*,[7] the Supreme Court considered the issue of sexual harassment for the first time. Neither party disputed that quid pro quo sexual harassment was impermissible under Title VII. Instead, the dispute in *Meritor* centered on whether a plaintiff should be entitled to relief if the plaintiff suffered no economic damage. Did the plaintiff need to show a "tangible loss of an economic character" to be protected by Title VII, or could a plaintiff's claim include noneconomic or psychological damage? In other words, did Mechelle Vinson have to prove she suffered some financial consequences because of the harassment in order to recover, or was her psychological pain enough to make her a winner?

With Justice Rehnquist penning the decision, the Court held that Title VII is not limited to just "economic" or "tangible" discrimination, but that harassment causing noneconomic injuries also can violate the Civil Rights Act. Agreeing with the EEOC, the Court declared that a claim of hostile environment sexual harassment is a form of sex discrimination that is actionable under Title VII. For the first time, hostile environment harassment or sexual conduct that has the purpose or effect of unreasonably interfering with an individual's work performance or creating an intimidating, hostile, or offensive working environment was judicially sanctioned by the highest court of the land as impermissible.[8]

But Who's Liable?

Along with the creation of hostile environment discrimination came the question of under what circumstances an employer should be liable for such harassment. Where an employee suffers an economic injury (loss of job, demotion, loss of benefits, etc.) because of a supervisor's actions, courts have had little difficulty holding an employer strictly liable based on agency principles.[9] That's because only someone with authority can fire, demote, or change an employee's benefits. A hostile environment, however, can be caused by supervisors or co-employees. Should an employer be strictly liable for harassment where a supervisor creates a discriminatory work environment, but does not make a job decision affecting the employee? The *Meritor* Court declined to consider this issue of employer liability, although it did note that an employer should not be "automatically" liable for harassment by a supervisor who creates a hostile environment.

Following the *Meritor* decision, the use of the terms quid pro quo and hostile environment took on an added meaning with respect to employer liability. And the standard of employer responsibility turned on which type of harassment had occurred. For example, if the plaintiff established a quid pro quo claim, the employer was subject to strict liability, even where the employer had no knowledge of the harassment and even if the employer had expressly forbidden the harassment. On the other hand, in hostile environment cases, the employer was liable only if it *knew or should have known* about the conduct and failed to correct it. Once the employer was "on notice," prompt effective action was required.[10] Still, the question of who is liable for supervisory harassment with no economic detriment (i.e., "Sleep with me . . . " but no firing) remained unanswered by the Supreme Court.

Now, for Faragher *and* Ellerth: *The Carrot and the Stick—The Gift*

Finally, twelve years after *Meritor,* in the companion cases of *Faragher v. City of Boca Raton*[11] and *Burlington Industries Inc. v. Ellerth,*[12] the Supreme Court squarely confronted the issue of employer liability in supervisor-created sexual harassment cases. The Court initially observed that for purposes of imposing liability on an employer, the sexual harassment categories of hostile environment and quid pro quo would no longer control.[13]

Analyzing the question of employer liability for the first time, the Court held that to impose "strict" or "no excuses" liability on an employer, something more than the employment relationship itself must be involved. Because, in a sense, most workplace "bad actors," whether supervisors or co-workers, are aided in accomplishing their objective through an employment relationship. That "something more," according to the Court, is a tangible employment action. The Court defined a tangible employment action as "a significant change in employment status, such as hiring, firing, failing to promote, reassignment with significantly different responsibilities, or a decision causing a significant change in benefits."[14] Similar to the end result of quid pro quo harassment, in most cases, a tangible employment action inflicts direct economic harm on the employee. Thus, when harassment results in a tangible employment action, the employer is strictly liable. End of story.

But what to do with supervisory harassment that does *not* result in a tangible employment action? Emphasizing "Title VII's basic policy of preventing violations by encouraging forethought and prophylactic measures by employers,"[15] the Court created the following rule:

> Employers are . . . liable for harassment committed by their supervisory employees. However, if the harassment does not culminate in a tangible employment action, an employer may raise the following affirmative defense and escape liability by showing:
>
> (a) that the employer exercised reasonable care to prevent and correct promptly any harassing behavior, and (b) that the plaintiff employee unreasonably failed to take advantage of any preventive or corrective opportunities provided by the employer or to avoid harm otherwise.[16]

This is called the "affirmative defense," which, if proven, can defeat liability in cases where the supervisor did not take a tangible employment action. It is the gift to employers to defeat liability.

Let's look at parts (a) and (b) of the affirmative defense in more detail because as an employer, you must "win" both parts:

(a) *that the employer exercised reasonable care to prevent and correct promptly any harassing behavior.* To fulfill the first prong of the affirmative defense, an employer has to use "reasonable care" to prevent harassment,[17] and has to use "reasonable care" to "correct promptly" any harassment that occurs. What might go into preventing harassment? Certainly, effective policies, training, reporting procedures, and enforcement will go a long way in preventing harassment. Correcting promptly? Enforcing the policy consistently and investigating promptly will address that issue. But what is "reasonable care"? There is no uniform answer. Yet, if an organization does not feel completely comfortable that its current procedures and actions are reasonable, they probably aren't reasonable. If so, an organization needs to do more, or may not have the opportunity to take advantage of the affirmative defense.[18]

(b) *that the plaintiff employee unreasonably failed to take advantage of any preventive or corrective opportunities provided by the employer or to avoid harm otherwise.* With respect to part (b) of the affirmative defense, the Court developed a corresponding obligation of "reasonable care" on the part of the complaining party. That person must be reasonable in taking advantage of the organization's policy, reporting procedures, and other corrective or preventive measures.

Thus, if an employer has a well-disseminated harassment prevention policy that includes a reasonable complaint procedure, has trained employees on the policies and procedures, and an employee fails to use it, the employer will likely "win" both parts (a) and (b), provided no tangible employment action was taken against the complaining employee. On the other hand, employers *without* disseminated sexual harassment policies and complaint procedures or complaint procedures that are unreasonable may be liable for harassing conduct by supervisors, regardless of whether the harassment results in a tangible employment action.

Kolstad: More Carrot, More Stick—Another Gift

In June of 1999, the Supreme Court handed down a third decision that will have long-lasting ramifications to the world of employment discrimination law and harassment, *Kolstad v. American Dental Association*.[19] In *Kolstad*, the Court held that an employer can be liable for additional damages, called punitive damages, under Title VII when the employer has acted "with malice or reckless indifference" to the federally protected rights of the employee.[20] The Court concluded that the term "malice or reckless indifference" pertains to the employer's knowledge that it may be acting in violation of federal law, not its awareness that it is engaging in discrimination.[21] Thus, an employer must discriminate in the face of a perceived risk that its actions will violate federal law to be liable for punitive damages.

However, the Court realized that since knowledge of federal law would now become a prerequisite for punitive damages, employers who in good faith educate their workforce about federal discrimination laws and Title VII should not be penalized with a risk of punitive damages. As it did in *Faragher* and *Ellerth*, the Court created a carrot for those who "do the right thing." The Court held that employers may not face the possibility of punitive damages for the discriminatory employment decisions of managerial and supervisory agents where the decisions are contrary to the employer's good-faith efforts to comply with Title VII.[22] This is a *Kolstad* defense, and the second gift to employers.

Similar to *Faragher* and *Ellerth*, the Court left to the lower courts the determination of what measures constitute "good faith efforts," stating only that "Title VII is designed to encourage the creation of anti-harassment policies and effective grievance mechanisms," and "the purposes underlying Title VII are similarly advanced where employers are encouraged to adopt anti-discrimination policies and to *educate* their personnel on Title VII's prohibitions" [emphasis added].[23]

What Has Happened Since Faragher/Ellerth/Kolstad: *The Beat Goes On*

With these pronouncements by the Supreme Court, lower courts and the EEOC were left to decipher and define the *Faragher/Ellerth* affirmative defense and the *Kolstad* "good faith" defense. What was patently clear, however, was that the defenses were part and parcel of the same court-mandated behavior—that employers would need to take affirmative steps to protect their workplace from harassment and discrimination and to shield themselves from liability and/or punitive damages. The policies and procedures of an employer thus became instrumental in preventing legal exposure. But what actual steps need be taken to protect an organization from liability and punitive damages? The answers grew more clear as the case law developed over the next several years.

Living and Dying by the First Prong of the *Faragher/Ellerth* Affirmative Defense

("that the employer exercised reasonable care to prevent and correct promptly any harassing behavior").

While the creation and distribution of a harassment prevention policy goes a long way toward meeting the first prong of the *Faragher/Ellerth* affirmative defense,[24] the mere existence of such a policy floating somewhere in the organization will likely prove insufficient to establish it.[25] Courts have held that simply forcing employees to sign a policy or merely issuing a policy does not constitute reasonable care.[26] For example, an employer's lack of knowledge about a policy and its procedures may prevent an employer from establishing part (a) of the affirmative defense.[27] In addition, the lack of a proper reporting mechanism providing alternative avenues for lodging complaints may also doom a properly distributed policy.[28] Even a perfectly drafted and disseminated policy may not insulate an employer from liability.[29] An employer must ensure that persons designated to receive complaints are adequately trained to respond to complaints.[30] Further, the employer must conduct a thorough, prompt, and effective investigation of the complaint and not retaliate.[31] In short, courts have reasoned that simply drafting a policy is not enough to safely meet the first prong of the affirmative defense; the employer must effectively implement, abide by, and live up to the policy's standards.

TIPS:
- Look at your policy. To whom do reports of harassment go? Are they reasonably accessible to receive complaints? Has everyone been trained on what they should do both in receiving complaints and in directing complaints to the appropriate persons to investigate?
- Do the individuals in the reporting procedure know never to ignore a complaint because the complaining party asked for secrecy (or an off-the-record conversation) or failed to use "magic" words like "I have a sexual harassment complaint"?
- Can the managers and supervisors recognize harassment if they see it, hear it, or hear about it? In our experience, many act only when someone actually uses the "magic" words (or files a form in triplicate)—*not*, as is necessary, when they learn about it through other means (even if it is by others who are not directly involved).

Living and Dying by the Second Prong of the *Faragher/Ellerth* Affirmative Defense

(*"that the plaintiff employee unreasonably failed to take advantage of any preventive or corrective opportunities provided by the employer or to avoid harm otherwise"*).

The second prong of the *Faragher/Ellerth* affirmative defense is based on the reasonableness of the employee's conduct—did the employee fail to take advantage of the employer's policies and procedures? Courts have looked primarily to whether the employee complained at all,[32] delayed in making a complaint,[33] or made the complaint according to established procedures.[34] Courts also will consider whether the employee may have been justified in not using an employer's policies and complaint procedures, most likely because of a reasonable fear of retaliation.[35] Finally, courts look to whether the employee reasonably failed to avoid harm.[36]

The reasonableness of an employee's actions in reporting harassment will certainly be dependent on the employer's attitude toward the harassment prevention policy.[37] The less attention given to the policy by the employer, the more likely an employee's confidence will be lessened in using the policy's reporting procedures. A poor work environment that fosters retaliation and gives employees a reason not to complain will also undermine the employer's affirmative defense. Such an environment will make an employee's failure to use a complaint procedure more reasonable and the employer liable, despite the possible existence of a well-disseminated policy and complaint procedure. Yet an employer who treats the policy as a critical component of its operations will relay the message that its harassment prevention policy and the procedures that accompany it are no less important than other rules and procedures, and thus must be followed.

TIPS: ■ Is your policy disseminated widely and often?
■ Can you show by an employee's signature whether he or she has received the policy?
■ Are there multiple avenues of complaint listed in the policy?
■ Is there a statement in your policy prohibiting retaliation?
■ Do you perform a fair and thorough investigation of all complaints?
■ Is your policy championed by management?

Living and Dying by *Kolstad*'s Good Faith Defense

In *Kolstad*, the Supreme Court indicated that an anti-discrimination policy can prevent an employer from being held liable for punitive damages. However, like the affirmative defense of *Faragher* and *Ellerth,* simply drafting such a policy does not automatically bar the imposition of punitive damages.[38]

In *EEOC v. Wal-Mart Stores, Inc.,*[39] an appellate court held that "the extent to which an employer has adopted anti-discrimination policies *and educated its employees* about the requirements of [Title VII] is important in deciding whether it is insulated from vicarious punitive liability" [emphasis added].[40] Cautioning that having a written policy is not enough, the court stated that "a generalized policy of equality and respect for the individual does not demonstrate an implemented good faith policy of *educating* employees" [emphasis added].[41] Perhaps the most strongly worded decision to date comes from the 7th Circuit case, *Mathis v. Phillips Chevrolet,*[42] where the court admonished, "leaving managers with hiring authority in ignorance of the basic features of the discrimination laws is an 'extraordinary mistake' . . . that can amount to reckless indifference."[43]

A court will generally look at the totality of the circumstances in determining liability for punitive damages. From a sound complaint procedure that gives the complainant many options to deliver his/her complaint and the ability to complain to someone who is not in his/her chain of command, to the quality of the investigation to the active dissemination of a policy, all of these will help to establish "good faith."[44]

TIPS: ■ Drafting a policy is not enough—Train. Train. Train.

The EEOC (Equal Employment Opportunity Commission)

Mirroring case law, the EEOC also takes a totality of circumstances approach when determining whether an employer should be liable for harassment. The regulations issued by the EEOC on sexual harassment provide that "[a]n employer should take all steps necessary to prevent sexual harassment from occurring, such as affirmatively raising the subject, expressing strong disapproval, developing appropriate sanctions, informing employees of their right to raise and how to raise the issue of harassment under Title VII, and developing methods to sensitize all concerned."[45]

And, shortly after the Supreme Court's decision in *Faragher/Ellerth*, the EEOC issued its post–*Faragher/Ellerth* Enforcement Guidance in 1999. The EEOC Enforcement Guidance provides that to exercise reasonable care under the first prong of the affirmative defense, an employer should establish, disseminate, and enforce an anti-harassment policy and complaint procedure.[46] Significantly, the EEOC Enforcement Guidance points out that "there are no . . . 'safe harbors' for employers based on the written content of policies and procedures. Even the best policy and complaint procedure will not alone satisfy the burden of proving reasonable care if, in the particular circumstances of a claim, the employer failed to implement its process effectively."[47]

I Have a Policy—Do I Have to Train, Too?

Based on case law and the EEOC, it is readily apparent that the mere creation and distribution of a harassment prevention policy is not enough to fully protect an employer from liability or punitive damages. Case law provides that courts will look to a number of factors to determine an employer's liability and potential for punitive awards. What we also know is that an employer must make an effort to ensure that the policy is understood and followed. Without full comprehension of the policy, along with the perception that the employer takes the policy seriously and will abide by its rules, an employer will continue to risk liability and punitive damages. Thus, an employer should ensure that managers and supervisors understand their responsibilities under the organization's anti-harassment policy and complaint procedure. According to the EEOC, one way to achieve that objective is to conduct periodic training of those individuals. In fact, the EEOC asserts that "if feasible, the employer should provide training to all employees to ensure they understand their rights and responsibilities."[48] While training may not be absolutely necessary to insulate an employer from liability and punitive damages in all instances, it certainly will provide a valuable insurance policy against such possibilities.

And, while *Faragher* and *Ellerth* did not address the impact of sexual harassment training on meeting the affirmative defense, several courts have relied on an employer's sexual harassment training to find that the employer did or did not exercise reasonable care to prevent harassment.[49] One court even stated that the gravamen of an "effective" harassment prevention policy includes training for its supervisors regarding sexual harassment.[50]

Another court found, as a matter of law, that lack of a harassment prevention policy at the time of the harassment and failure to conduct training until more than one year after the offender admitted the harassment, and more than two and a half years after the harassment first began, constituted lack of reasonable care.[51] In another case, a court observed that the company did not exercise reasonable care where training wasn't held at a particular site, observing, "[B]ecause having its employees be aware of the policy is so crucial to having a policy that is effective, and based on the evidence presented at trial, it is seriously doubtful that [the company] could be said to have 'exercised reasonable care to prevent and correct promptly any sexually harassing behavior.'"[52]

Clearly, harassment training is a determinant in proving part (a) of the affirmative defense—the employer's exercise of reasonable care. However, harassment training can also be relevant to meet part (b) of the defense because proper training can make an employee's failure to use complaint procedures more unreasonable. Also, with respect to co-worker harassment or other forms of harassment, training a workforce has become a crucial factor in eliminating employer liability.[53]

Training has also proved to be critical in establishing the good faith defense of *Kolstad*.[54] For example, one court held that to insulate itself from punitive damages, an employer must show efforts have been made to implement its anti-discrimination policy through education of its employees, active enforcement of its mandate, training of its supervisors, and periodic dissemination of the policy.[55]

Meeting *Kolstad*'s good faith defense is imperative not only to minimize liability, but also to minimize the chance of extended litigation. Remember, engaging in quality training in conjunction with a well-written policy will likely translate into successfully meeting the good faith defense of *Kolstad*. Thus, even if an employer loses summary judgment on the merits and must go to trial, the ability of the employer to show that it made "good faith efforts" to comply with Title VII can result in the punitive damage claim being dismissed. When punitive damages are no longer an available remedy, the value of the case drops significantly, and the case is more likely to be quickly resolved.

With the monetary value of a lawsuit now often hinging on the effectiveness of an employer's preventive policies, there has been a new emphasis during discovery on an employer's "reasonable care" and "good faith efforts" to prevent harassment/discrimination, or unfortunately, sometimes, an employer's lack thereof.[56] Plaintiff attorneys and the EEOC have begun to aggressively question employers on how much money is spent on training, the expertise of the trainers, the curriculum, and employee response to the training.[57] Thus, as training programs have become increasingly important, the quality of these programs has developed into the newest battlefield in the employment litigation wars.

State Law Requirements

While the need for training is apparent under federal law, training is also a necessary tool for protection under state law.[58] In fact, required training is here in many states and on the way in others. While most of those states require it of public employees, a few states have extended the requirement to private employers.[59] And it will only be a matter of time before the requirement of sexual harassment training is extended to other forms of harassment. Employers should have counsel review relevant state laws to ensure compliance.

Okay, So I'll Train, but How Often and What Kind?

While the Supreme Court did not spell it out for us, they did tell us we have to use reasonable care. Does that mean we have to train everybody every day? That seems very unreasonable to us. How about every decade? Swinging the pendulum so far also seems unreasonable. Many employers (in states where there are no legal requirements respecting training) look at their employee turnover rate, internal claim rate, EEOC charges, how centralized or decentralized the organization is, and the confidence level of the organization in successfully handling complaints to determine how often they should train employees. Obviously, the greater the concerns, the more frequent the need to train.

But what kind of training must be delivered each time? Again, the Supreme Court was silent on the details, but there are numerous delivery methods that can be part of any organization's ongoing training initiative, including leader-led training, Web-based training, self-studies, and staff meetings. Don't forget that new employee orientation videos may be very useful to "inoculate" your new employees from the time they come to work on their first day until they are trained on the policy and the related content.

Some organizations support the training by distributing the policy on Friday afternoons before happy hour, as a gentle reminder about off-premises conduct. Still other organizations put the policy in envelopes with paychecks or direct deposit statements as a way of supporting the training. We have seen some employers print their reporting procedures on wallet cards and mouse pads to supplement the routine distribution of their entire policy. Certainly these ideas and procedures couldn't hurt and might even help prove you have met the reasonable-care standard.

It is more important than ever for employers to implement quality harassment training programs. Structured properly, employee training not only mitigates potential liability and helps eliminate punitive damage awards, but also adds tremendous value to an organization and can eliminate problems of harassment and discrimination before they rise to litigation issues.

This, of course, begs the question of how one implements a quality harassment training program. The answer to this question lies in the chapters ahead.

LESSONS LEARNED

To: Millionaire
From: The Trainers
Re: Lessons Learned

Dear Millionaire:

It is so important to know about the *Faragher/Ellerth* and *Kolstad* defenses—they have big implications for our harassment training. Here are some tips on how to best ensure you receive your "gifts" from the Supreme Court:

- It is important to focus on your harassment prevention policy in training.
- If your managers and supervisors are involved in quid pro quo harassment, you are liable whether you knew about their conduct or not. No excuses permitted, write the check. If, in an imperfect world, you cannot train everyone, focus on your managers and supervisors. They have the ability to get you into a whole lot of trouble.
- If you train your employees, managers, and supervisors on the conduct that your policy prohibits, you will increase your chances of succeeding on the first prong of the affirmative defense.
- If you train employees on your own reporting policy and have effective reporting procedures, investigations, and a non-retaliatory system, you are likely to succeed on the second prong of the affirmative defense.
- Make sure you train all of your workforce, including your upper-level executives. They must understand what the workplace "rules" are.

- The burden of proving the affirmative defense and the *Kolstad* defense is on the organization. Therefore it is imperative to document and retain everything that may be part of that defense.

Signed,

The Trainers

Notes

1. Remember the importance of avoiding punitive damages. Yes, if you get to the question of punitive damages, it means you lost the case on liability. But the big numbers found in jury verdicts are usually the result of exorbitant punitive damages. *See* Swinton v. Potomac Corp., 270 F.3d 794 (9th Cir. 2001) (in a racial harassment suit, the jury awarded $30,000 for emotional distress, $5,612 for back pay, and $1 million in punitive damages); Thorne v. Sprint Communications Co., No. 00-00913-HRS (W.D. Mo. March 29, 2002) (in a sexual harassment suit, a federal jury awarded $1.1 million in punitive damages, $100,000 for pain and suffering, and a $2,850 bonus, which the plaintiff failed to receive). *See also,* Carroll v. Interstate Brands Co., Calif. Super. Ct., No. 995728 (award August 2, 2000) (San Francisco jury awards seventeen former and current black bakery workers $120 million in punitive damages on their claims of race discrimination).

2. *See infra* notes 10 through 54.

3. Section 703(a): Title VII of the Civil Rights Act of 1964, 42 U.S.C. § 2000e [§ 703(a)].

4. 29 C.F.R. § 1604.11(f).

5. In a classic quid pro quo case, the issue is whether an individual has suffered a job detriment for refusing to accede to unwelcome advances. In the classic case, the issue is not *whether* an adverse employment action occurred, but rather *why* it occurred. The complainant argues that a substantial motivation for an adverse employment action was the complainant's refusal to succumb to a sexual advance. The defense typically is that the job detriment reflected a nondiscriminatory business reason, that there was no sexual advance, or that the sexual advance was welcomed.

6. In a hostile environment case, the issue is whether incidents of unwelcome conduct based on gender have been so "severe" or "pervasive" that they (1) subjectively altered the conditions of the complainant's employment and (2) would have altered the conditions of the employment of a reasonable person in the complainant's position. Incidents that are relatively trivial or isolated do not create liability.

7. 477 U.S. 57 (1986).

8. Seven years later, the Supreme Court discussed the factors needed to prove that a hostile environment exists in Harris v. Forklift Systems, Inc., 510 U.S. 17 (1993). The Harris Court stated several basic principles:

- Whether a hostile environment exists depends upon a combination of circumstances—the frequency of discriminatory conduct; its severity; whether it is physically threatening or humiliating; whether it unreasonably interferes with an employee's work performance; and/or whether it causes psychological injury.
- No one factor is required to find liability. For example, a hostile environment can exist even in the absence of any psychological injury.

■ To be hostile, the environment must be subjectively perceived as abusive by the complainant; a thick-skinned complainant who is not really bothered by even outrageous conduct has no legal claim.

■ A finding of hostile environment also requires that the environment was objectively abusive in the sense that it would have been abusive to a "reasonable person"; a thin-skinned complainant who is hypersensitive to conduct that would not affect reasonable people would have no legal claim.

9. "Strict" liability, in this sense, means "no excuses" liability. It doesn't matter whether you knew about the harassment or not, you are going to be liable no matter what.

10. With respect to co-worker harassment, there has been little debate as to the liability standard, and the following rule has been uniformly applied by courts. An employer is liable for co-worker acts of sexual harassment in the workplace where the employer (or its agents or supervisory employees) knows or should have known of the conduct, unless it can show that it took immediate and appropriate corrective action. Actual notice may be found if an employee complained to a member of management. Constructive notice may be found if the harassment is so pervasive that a reasonable employer would have discovered the facts. This is known as the "negligence standard."

11. 524 U.S. 775 (1998). *See* Appendix A for a copy of the case.

12. 524 U.S. 742 (1998). *See* Appendix B for a copy of the case.

13. In reaching its holding, the Court observed that most lower courts began their analysis of sexual harassment claims by categorizing the alleged conduct as either quid pro quo or hostile environment sexual harassment. Dismissing their usefulness in assessing employer liability, the Court opined that the categories are useful only as a "rough demarcation between cases in which threats are carried out [quid pro quo] and those where they are not or are absent altogether [hostile environment]." *Burlington* at 751.

14. *Burlington* at 761.

15. *Id.* at 764.

16. *Id.* at 765.

17. An employer's failure to have a harassment prevention policy, while not dispositive, appears to be one of the major factors in deciding whether (a) was met. *Id.*

18. *See infra* discussion.

19. 527 U.S. 526 (1999). *See* Appendix C for a copy of the case.

20. *Id.* at 535.

21. *Id.*

22. *Id.* at 545–46.

23. *Id.* at 545.

24. Barrett v. Applied Radiant Energy Corp., 240 F.3d 262 (4th Cir. 2001) (distribution of an anti-harassment policy provides compelling proof that the company exercised reasonable care in preventing and properly correcting sexual harassment. The court noted that the only way to rebut this proof is to show that the policy was defective or dysfunctional or was adopted or administered in bad faith).

25. In *Faragher*, the Court held that even though the city had a sexual harassment policy, it was liable for the harassment because the policy was not distributed to all employees. 524 U.S. at 808–9.

26. *See* Molnar v. Booth, 229 F.3d 593 (7th Cir. 2000) (affirmative defense failed since the employer had only a general anti-discrimination policy and no sexual harass-

ment policy, and employees were "extremely confused" about what sexual harassment actually was); Frederick v. Sprint/United Management Co., 246 F.3d 1305 (11th Cir. 2001) (employer failed to establish that its policy was effectively published or contained reasonable complaint procedures); Smith v. First Union Nat'l Bank, 202 F.3d 234 (4th Cir. 2000) (court held that an employer cannot establish an affirmative defense as a matter of law with a "defective or dysfunctional" policy that implied only sexual advances constitute sexual harassment).

27. *See, e.g.,* Lancaster v. Scheffler Enterprises, 19 F. Supp. 2d 1000 (W.D. Mo. 1998) (employer must take reasonable steps to prevent, correct, and enforce policy).

28. *See* Ocheltree v. Scollon Prods., Inc., 161 F.3d 3 (4th Cir. 1998) (an open-door policy, although informal and without any mention of sexual harassment, may also be dispositive of reasonable care to correct problems. But an *ineffective* open-door policy will not satisfy the first prong, for example, if managers and supervisors were unavailable/unresponsive to employees who tried to use the open-door policy); Leslie v. United Technologies Corp., 51 F. Supp. 2d 1332 (S.D. Fla. 1998) (defendant cannot escape liability by merely adopting ineffective diversity and grievance mechanisms). *But see* Madray v. Publix Supermarkets, Inc., 208 F.3d 1290 (11th Cir. 2000) (sexual harassment policy provided alternative avenues for lodging complaints, and although only one person in the store was always present to receive complaint, others were available, as was a toll-free telephone number). *See also* EEOC Enforcement Guidance: Vicarious Employer Liability for Unlawful Harassment by Supervisors (6/18/99), EEOC Compliance Manual (BNA) (EEOC Enforcement Guidance) (a complaint process is not effective if employees are always required to complain first to their supervisor about alleged harassment. . . . It is advisable for an employer to designate at least one official outside an employee's chain of command to take complaints of harassment).

29. *See, e.g.,* Williams v. Spartan Communications, Inc., 210 F.3d 364 (4th Cir. 2000) (Table) (outrageous comments and behavior by upper management may provide evidence that an anti-harassment policy is ineffective).

30. *See* Smith v. First Union Nat'l Bank, 202 F.3d 234 (4th Cir. 2000) (court rejected the employer's affirmative defense based on the fact that the employer's investigator had never done one before, never asked the alleged harasser whether he made comments alleged, and simply counseled him to improve his management style and "smile more"); Williams v. Spartan Communications, Inc., 210 F.3d 364 (4th Cir. 2000) (Table) (supervisor received no sexual harassment training and could not even recall any discussions about the sexual harassment policy); Gentry v. Export Packaging Co., 238 F.3d 842 (7th Cir. 2001) (court concluded that even though the company maintained a clear harassment prevention policy, managers and supervisors unclear about who was to take harassment complaints, in part, rendered it liable).

31. *See* New Hampshire Dep't of Corr. v. Butland, N.H., No. 2000-803 (May 7, 2002) (initiating an investigation within hours of the report and suspending the alleged harasser by the next day supported a finding that a state agency acted promptly and should avoid liability); Hill v. American General Fin. Inc., 218 F.3d 639 (7th Cir. 2000) (even though the harassment policy was not individually distributed and simply forbade discrimination, court emphasized that the employer responded to the employee's first non-anonymous letter by promptly investigating, reprimanding harasser, and transferring supervisors—the employer's harassment policies also were easily available for employee review and included a complaint procedure and telephone numbers); Stuart v. GMC, 217 F.3d 621 (8th Cir. 2000) (investigated promptly, redistributed sexual harassment policy, and offered the complainant

transfer to different department—court also noted the response was reasonable when considering factors of (a) time elapsed between notice and response, (b) options available to employer, (c) disciplinary steps taken, and (d) whether response ended harassment); Casiano v. AT&T Corp., 213 F.3d 278 (5th Cir. 2000) (court affirmed summary judgment for the employer where the employer distributed a sexual harassment policy, its supervisors reviewed the policy with the plaintiff, and then promptly suspended the alleged harasser after fully investigating the harassment report); Scrivner v. Socorro Ind. School Dist., 169 F.3d 969 (5th Cir. 1999) (employer's prompt response—including immediate investigation, warning to harasser, and termination of harasser after subsequent complaint—satisfied first prong); Corcoran v. Shoney's Colonial, 24 F. Supp. 2d 601 (W.D. Va. 1998) (immediately investigating harassment, minimizing the contact between the alleged harasser and the complaining employee, and confronting the harasser with the allegations, whereupon the alleged harasser resigned, was adequate to correct harassment); Fiscus v. Triumph Group Operations, 24 F. Supp. 2d 1229 (D. Kan. 1998) (investigating an incident and demoting the harassing supervisor upon his confession to the harassment, after which the supervisor resigned, was adequate to correct harassment); Hubbard v. UPS, 200 F.3d 556 (8th Cir. 2000) (court emphasized that the employer promptly explained the sexual harassment policy to the harasser, warned him, transferred him to a new area, and offered the complainant her own transfer); Johnson v. West, 218 F.3d 725 (7th Cir. 2000) (court upheld the trial court, finding that the employer exercised reasonable care to prevent and correct harassment where the employer had an established harassment policy, immediately investigated the plaintiff's report of harassment, and separated the plaintiff and the alleged harasser—court noted that the employer even continued to investigate when the plaintiff asked the employer to stop the investigation); Reinhold v. Virginia, 151 F.3d 172 (4th Cir. 1998) (absence of further harassment evidences effectiveness of employer's response). *But see*, Howley v. Town of Stratford, 217 F.3d 141 (2d Cir. 2000) (that employer took five weeks to discipline the harasser, issued only a weekend suspension, and told the harasser to apologize to the complainant, which he did not do, leading to employer liability).

32. *See, e.g.,* Coates v. Sundor Brands, Inc., 164 F.3d 1361 (11th Cir. 1999) (employee's failure to use formal complaint mechanism set forth in the harassment policy and her choice to instead make informal statements to co-workers, which were passed on to management anonymously, and to make ambiguous statements to supervisors not specifying that sexual harassment was in question, amounted to a failure to adequately notify the company of the harassment).

33. *See* Greene v. Dalton, 164 F.3d 671, 675 (D.C. Cir. 1999) (to defeat the second prong, employer must show more than an excusable delay in reporting the harassment. It must additionally show that "a reasonable person in [plaintiff's] place would have come forward early enough to prevent [the] harassment from becoming severe or pervasive."); Phillips v. Taco Bell Corp., 156 F.3d 884 (8th Cir. 1998) (whether plaintiff's three-month delay in reporting harassment was reasonable is a question of fact for the jury); Watts v. Kroger Co., 170 F.3d 505 (5th Cir. 1999) (plaintiff's three- to four-month delay between harassment and complaint was not unreasonable); Corcoran v. Shoney's Colonial, 24 F. Supp. 2d 601 (W.D. Va. 1998) (plaintiff's eight-month delay in reporting harassment was not unreasonable); Fall v. Indiana University Board of Trustees, 12 F. Supp. 2d 870 (N.D. Ind. 1998) (three-month delay in reporting harassment was reasonable because final reporting was accompanied by an increase in the intensity of the harassment); Scrivner v. Socorro

Ind. School Dist., 169 F.3d 969 (5th Cir. 1999) (plaintiff's eight- to nine-month delay between harassment and complaint was unreasonable); Montero v. AGCO Corp., 192 F.3d 856 (9th Cir. 1999) (two-year delay was unreasonable); Dedner v. Oklahoma, 42 F. Supp. 2d 1254 (E.D. Okla. 1999) (three-month delay was unreasonable); Speight v. Albano Cleaners, Inc., 21 F. Supp. 2d 560 (E.D. Va. 1998) (plaintiff's failure to complain until the day she quit was unreasonable). *See also* EEOC Enforcement Guidance: Vicarious Employer Liability for Unlawful Harassment by Supervisors (6/18/99), EEOC Compliance Manual (BNA) (EEOC Enforcement Guidance) (an employee might reasonably ignore a small number of incidents, hoping that the harassment would stop without resort to complaint).

34. *See* Anderson v. Deluxe Homes of Pennsylvania, Inc., 131 F. Supp. 2d 637 (M.D. Pa. 2001) (even though employee complained to team manager and supervisor rather than management, jury should decide whether the plaintiff reasonably believed that her team manager and supervisor were in a position to either stop the harassment or inform higher management of the problem); Maddin v. GTE, Inc., 33 F. Supp. 2d 1027 (M.D. Fla. 1999) (complaining to a union steward was not reasonable notice to the employer where the sexual harassment policy clearly listed a number of entities to which complaints of sexual harassment should be made). *But see* EEOC Enforcement Guidance (an employee does not unreasonably fail to complain if the employee boycotts the employer's grievance system and instead relies on the EEOC or a union grievance); Watts v. Kroger Co., 170 F.3d 505 (5th Cir. 1999) (filing a union grievance instead of using employer's sexual harassment policies satisfies second prong); Breda v. Wolf Camera & Video, 222 F.3d 886 (11th Cir. 2000), complaint dismissed, 148 F. Supp. 2d 1371 (S.D. Ga. 2001) (court held that an employee who follows a policy need not be concerned about pursuing a complaint further up the employer ladder; instead, the sole inquiry when the employer has a clear and published policy is whether the complaining employee followed the procedures established in the policy).

35. *See* Thomas v. BET Soundstage Restaurant, 104 F. Supp. 2d 558 (D. Md. 2000) (question of fact as to employee's reasonableness where she knew about reporting procedure of policy, but evidence suggested that her manager discouraged complaints and her supervisor was infamous for firing people); Katt v. City of New York, 151 F. Supp. 2d 313 (S.D.N.Y. 2001) (issue as to whether police officer feared retaliation for reporting harassment). *See also* Anderson v. Deluxe Homes of Pennsylvania, Inc., 131 F. Supp. 2d 637 (M.D. Pa. 2001).

36. *See, e.g.,* Brown v. Perry, 184 F.3d 388 (4th Cir. 1999) (federal agency was not vicariously liable for a manager's sexual harassment of a subordinate who went to the harassing manager's hotel room after a night of bar hopping only a few months after she was allegedly harassed by the same manager (under similar circumstances the company had previously disciplined the manager), because, in light of the previous history between the manager and the "victim," the plaintiff utterly failed to avoid harm).

37. Because a policy must prohibit not only sexual harassment, but harassment based on other protected characteristics, a policy is correctly a "harassment prevention policy" or "anti-harassment policy" rather than a "sexual harassment policy."

38. *See, e.g.,* Bruso v. United Airlines, Inc., 239 F.3d 848 (7th Cir. 2001) (although the implementation of a written or formal anti-discrimination policy is relevant, it is not sufficient in and of itself to insulate an employer from punitive damages awards. "If mere implementation were sufficient, employers would have an

incentive to adopt formal policies in order to escape liability for punitive damages, but they would have no incentive to enforce those policies").

39. 187 F.3d 1241 (10th Cir. 1999).

40. *Id.* at 1248–49.

41. *Id.*

42. 269 F.3d 771 (7th Cir. 2001).

43. *Id.* at 778.

44. *See* Deffenbaugh-Williams v. Wal-Mart Stores, Inc., 188 F.3d 278 (5th Cir. 1999) (a good faith defense was not established by evidence that employer encouraged employees to contact higher management with grievances; there was an absence of evidence that the policy was enforced or specific evidence that the plaintiff's complaints were investigated); Knowlton v. Teltrust Phones, Inc., 189 F.3d 1177 (10th Cir. 1999) (remanding the case for trial on the grounds that the plaintiff had presented enough evidence for a court to consider the punitive damages complaint, because the company knew of the harasser's conduct, otherwise expressed hostility toward women, and was unresponsive to the plaintiff's continued concern upon learning that the alleged harasser might still interact with her); Cadena v. Pacesetter Corp., 224 F.3d 1203 (10th Cir. 2000) (employer may have not made good faith efforts where manager and supervisor responsible for sexual harassment training admitted ignorance about sexual harassment); Blackmon v. Pinkerton Sec. & Investigative Servs., 182 F.3d 629 (8th Cir. 1999) (where good faith defense was not available because the investigation of plaintiff's complaint was "inadequate and disproportionate to the seriousness of [her] complaints"); Kimbrough v. Loma Linda Dev., Inc., 183 F.3d 782 (8th Cir. 1999) (punitive damages awarded where harasser's supervisor knew of the harassment but failed to act); Fuller v. Caterpillar Inc., 124 F. Supp. 2d 610 (N.D. Ill. 2000) (good faith was met where the policy was posted at building entrances, two booklets were provided to employees on recognizing and reporting sexual harassment, employer required employees to attend diversity training, and all employees were required to attend sexual harassment training); Carr v. Caterpillar, Inc., et al., No. 96-1485 (C.D. Ill. Sept. 30, 1999) (court sustained summary judgment for the defendants on the punitive damages claim where the employer had a written harassment prevention policy that was required to be posted in all offices of one co-defendant; attempts were made to educate all employees about sexual harassment through several training seminars at the same co-defendant's office; sexual harassment training was provided for managers and supervisors; and a mandatory supervisor training course included sexual harassment training at the other corporate co-defendant). *See also* EEOC Enforcement Guidance (a complaint process is not effective if employees are always required to complain first to their supervisor about alleged harassment. . . . It is advisable for an employer to designate at least one official outside an employee's chain of command to take complaints of harassment).

45. 29 C.F.R. § 1604.11(f).

46. EEOC Enforcement Guidance: Vicarious Employer Liability for Unlawful Harassment by Supervisors (6/18/99), EEOC Compliance Manual (BNA) (EEOC Enforcement Guidance). The EEOC Enforcement Guidance also provides that the rule in *Faragher* and *Ellerth* regarding vicarious liability applies to harassment by supervisors based on race, color, sex (whether or not of a sexual nature), religion, national origin, age, or disability. Thus, employers should establish anti-harassment policies and complaint procedures covering *all* forms of unlawful harassment. *See*

Appendix D for a complete copy of the Guidance. For a complete copy of the EEOC Guidance on the Web, go to http://www.eeoc.gov.

47. *Id.*

48. *Id.*

49. *See* Shaw v. Auto Zone, Inc., 180 F.3d 806 (7th Cir. 1999) (distributing policy and regularly conducting training established reasonable care); Richardson v. New York State Department of Correctional Services, 180 F.3d 426 (2d Cir. 1999) (lack of training leads to failure of affirmative defense); Burrell v. Crown Central Petroleum, Inc., 121 F. Supp. 2d 1076 (D. Tex. 2000) (training led to meeting affirmative defense); Thomas v. BET Soundstage Restaurant, 104 F. Supp. 2d 558 (D. Md. 2000) (training was a factor in affirmative defense); Miller v. Woodharbor Molding & Millworks, Inc., 80 F. Supp. 2d 1026 (N.D. Iowa 2000), aff'd by 248 F.3d 1165 (8th Cir. 2001) (failure to educate managers and supervisors about sexual harassment policies defeated first prong of affirmative defense); Wal-Mart Stores v. Davis, 979 S.W.2d 30 (Tex. App. Austin 1998) (failure to educate managers and supervisors about sexual harassment policies defeated first prong of affirmative defense); Elmasry v. Veith 2000 WL 1466104 (D.N.H. January 2000) (genuine issue of material fact as to whether the mere distribution of a handbook, without proper training, is effective to prevent sexual harassment from taking place). *See also* Anderson v. Leigh, 2000 WL 193075 (N.D. Ill. February 10, 2000); Romero v. Caribbean Restaurants, 14 F. Supp. 2d 185 (D.P.R. 1998); Masson v. School Board of Dade County, 36 F. Supp. 2d 1354 (S.D. Fla. 1999); Maddin v. GTE, 33 F. Supp. 2d 1027 (M.D. Fla. 1999); and Fiscus v. Triumph Group Operations, Inc., 24 F. Supp. 2d 1229 (D. Kan. 1998).

50. Miller v. Woodharbor Molding & Millworks, Inc., 80 F. Supp. 2d 1026 (N.D. Iowa 2000).

51. Hollis v. City of Buffalo, 28 F. Supp. 2d 812 (W.D.N.Y. 1998).

52. Nuri v. PRC, Inc., 13 F. Supp. 2d 1296 (M.D. Ala. 1998).

53. *See* Stuart v. General Motors Corp., 217 F.3d 621 (8th Cir. 2000) (training is also a factor in the negligence arena); Booker v. Budget Rent-A-Car Systems, 17 F. Supp. 2d 735 (M.D. Tenn. 1998) (training is a factor for meeting the affirmative defense for racial harassment).

54. *See* Cooke v. Stefani Management Services Inc., 250 F.3d 564 (7th Cir. 2001) (restaurant established good faith where it had a sexual harassment policy, harassing manager and supervisor attended a seminar on sexual harassment and the policy was posted at the restaurant where the plaintiff worked); Cadena v. Pacesetter Corp., 224 F.3d 1203 (10th Cir. 2000) (Wal-Mart recognized that to avail itself of *Kolstad*'s defense an employer must at least adopt anti-discrimination policies and make a good faith effort to educate its employees; quality of the training was relevant to determine whether the *Kolstad* defense was met); Anderson v. G.D.C. Inc., 281 F.3d 452 (4th Cir. 2002) (court noted that an employee could proceed with a claim for punitive damages because the employer did not have a policy on training or preventing discrimination); Marcano-Rivera v. Pueblo Int'l Inc., 232 F.3d 245 (1st Cir. 2000) (in ADA suit, court held that jury should not consider punitive damages because the employer had implemented a nondiscrimination policy and trained its employees on the ADA); Swinton v. Potomac Corp., 270 F.3d 794 (9th Cir. 2001) (affirming punitive damages award based in part on the company's failure to conduct anti-harassment training until nearly seven months after the plaintiff quit).

55. Romano v. U-Haul International, 233 F.3d 655 (1st Cir. 2000).

56. *See, e.g.,* Wolfe v. Village of Brice, 37 F. Supp. 2d 1021 (S.D. Ohio 1999) (questioning the details of defendant's sexual harassment training).

57. *See* Appendix E.

58. Whether states will determine liability based on the *Faragher/Ellerth* model is unclear. For example, at the time of this writing, the California Supreme Court is addressing whether the *Faragher/Ellerth* affirmative defense should be available for sexual harassment claims under the California Fair Employment and Housing Act. The Ninth Circuit predicted California's Supreme Court will adopt it. Kohler v. Inter-Tel Technologies, 244 F.3d 1167 (9th Cir. 2001). However, the Michigan Supreme Court refused to apply the U.S. Supreme Court's Faragher/Ellerth decisions to sexual harassment claims under Michigan's Civil Rights Act in Chambers v. Trettco, Inc., 614 N.W.2d 910 (Mich. 2000).

59. *See* Appendix F.

2

Not Another Policy: The Nuts and Bolts of Harassment Prevention "House Rules"

Dear Trainers:

We have an EEO policy that commits to a discrimination-free workplace and mentions sexual harassment. Isn't that enough? Do I have to have a twenty-pound harassment policy to keep from getting sued? Why can't employees just behave? It seems like people should just use their own common sense about sexual harassment, so why do we have to have a separate policy?

Signed,

Not Another Policy

. . . .

Dear Not Another Policy:

The heart of your legal defenses and training will be your policy. It is both your sword to fight off harassment and your shield from liability. We wish common sense were enough to guide employees' conduct, but everyone was raised with different values and rules. For example, in some homes, families swear and tell sexual (or racial, ethnic, disability, etc.) jokes while in others, such conduct is punished. Therefore, an organization needs to bring its employees together under one policy regarding conduct. We call the harassment prevention policy an organization's "house rules." Everyone grew up with house rules of some sort. Therefore, most employees readily understand your organization's need for house rules about behavior. However, for legal reasons, employers must go one step further than parents. Because the policy is unique to the workplace, employees need to be trained—to understand how to apply it when interacting with co-workers. The policy has to be clear, yet comprehensive. Even the Equal Employment Opportunity Commission has weighed in on what a solid policy should cover. Read on!

Signed,

The Trainers

Overview

As mentioned before, the U.S. Equal Employment Opportunity Commission issued an Enforcement Guidance regarding the Supreme Court decisions on sexual harassment.[1] Through its analysis of the 1998 Supreme Court decisions in *Burlington Industries, Inc. v. Ellerth* and *Faragher v. City of Boca Raton*, the EEOC's comprehensive policy guidance explains the circumstances under which employers can be held liable for unlawful harassment by supervisors. The Guidance addresses the steps employers should take to prevent and correct harassment and the nature of employees' obligations to bring complaints of harassment to their employers' attention.

We have designed this chapter to be a checklist so that you can either audit your current policy to make sure it is compliant with the EEOC or draft a new one. We have organized the Enforcement Guidance into sections that should correlate to your policy. This is not legal advice, nor is it intended to be legal advice, and we suggest you have an employment attorney review your final draft to make sure it is compliant, says what you mean to say,[2] and does not create unnecessary obligations.[3]

What You Need to Know . . .

Checklists

☑ **PROHIBITION AGAINST HARASSMENT**

- ○ *Give a clear explanation of prohibited conduct*
- ○ *Include race, color, religion, sex, national origin, age, disability, and all other protected characteristics*
- ○ *Include co-workers, managers, supervisors, and third parties*
- ○ *Encourage employees to report harassment before it becomes severe or pervasive*
- ○ *Assure that the organization will make every effort to stop harassment before it rises to the level of a violation of federal law*

TIPS:

- Consider making a clear "zero tolerance" statement about harassment so that there is no question about where your organization stands on the issue.
- Make sure to state that the list of prohibited conduct is not all-inclusive (or the policy *would* weigh twenty pounds).
- Remember to include protected categories based upon your state and local laws. State and local laws may differ greatly.
- If your organization is in more than one state, remember to address all protections. One way to do this may be to add to the end of your list, " . . . and all other state and local laws that apply." Many organizations pick the state that offers the most protections and tailor their nationwide policy to that state.
- You can protect other categories not protected federally or in your state. For example, many states do not prohibit harassment or discrimination based on sexual orientation. Many organizations, however, offer such protection in their policies even though they are not under any federal, state, or local obligation to do so.
- **Done wrong:** Be wishy-washy. Have a weak statement about harassment, such as "This organization prefers that employees do not harass each other." Believe it or not, we copied this from a policy we recently reviewed.

✔ EFFECTIVE COMPLAINT PROCESS

- ○ *Designed to encourage victims to come forward*
- ○ *Clearly explains the reporting process*
- ○ *Ensures no obstacles*
- ○ *Complaint does not have to be in writing*
- ○ *Must be accessible complaint process with at least one official outside of an employee's chain of command*
- ○ *Supervisors must be instructed to report complaints to appropriate officials*
- ○ *Includes time frames for filing with the EEOC or state Fair Employment Practice Acts (FEPA) and an explanation of the deadlines*

 TIPS:

- The complaint procedure is a critical item in your policy and your defense against liability. Draft it carefully!
- In addition to listing people in the reporting process by titles such as director of human resources, office manager, etc., consider also listing the names and direct telephone numbers of the people who hold those positions. This tells employees exactly whom they may go to. However, if you have a high turnover rate or run a large organization with many people in those functions, you may prefer not to include this information.[4] Also, if someone in the reporting procedure changes, an organization should revise its policy and redistribute it. This is another opportunity to show you are taking "reasonable care."

■ Hotlines. It's a great idea to have a 24-hour toll-free hotline where reports can be made at any time of the day or night, even anonymously (although you may want to have a statement in your policy that it may not be possible to resolve anonymous complaints).

■ If you list phone numbers, remember to actually call them to check that the numbers are accurate and free from typos. Don't forget to include area codes.

■ Many policies include a clear statement about not having to report harassment claims to the individual you are making the complaint about, but rather to another individual in the reporting procedure.

■ Make sure to train all those individuals listed in the reporting procedure in how to properly take in complaints so they do not make statements like, "Not this again. You are the biggest whiner."

■ Consider having both men and women in the reporting process to accommodate anyone who may feel more comfortable speaking with one or the other gender.

■ If you have multiple shifts, audit your procedure to make sure that individuals could also complain to someone who works that shift. For example, many employers include managers and supervisors in the reporting procedure because most hourly employees will have more face-to-face contact with those individuals than they would with the director of human resources. Having numerous alternatives in the reporting procedure is critical.

■ **Done wrong:** "All complaints must be submitted in writing to the Human Resources office at headquarters Monday through Friday between the hours of 9:00 a.m. and 3:00 p.m. Complaints must include a detailed description of the event with statements from witnesses . . ." You get the idea—avoid at all costs obstacles and roadblocks that may make the reporting procedure appear to be unreasonable to a judge or jury.

✔ CONFIDENTIALITY

○ *Provides confidentiality to the extent possible*
○ *Records are kept confidential*
○ *Supervisors must report all conduct that might violate the policy*
○ *Hotlines are good for anonymous calls*

TIPS:　■ Without a statement about confidentiality, individuals may fear reporting, thereby diminishing the effectiveness of your policy.

■ Individuals in the reporting procedure should be trained about keeping these matters confidential so that only those on a need-to-know basis will be involved.

■ **Done wrong:** A supervisor listed in the reporting process says to an employee making a complaint, "You know, this is going to get out to everyone and then you are going to have to live with the consequences of getting someone fired." Yes, indeed, we didn't make this up.

 EFFECTIVE INVESTIGATIVE PROCESS

○ *Provide prompt, thorough, and impartial investigation*
○ *Consider intermediate measures, but not the involuntary transfer of the complainant*
○ *Assure immediate and appropriate corrective action when harassment has occurred*
○ *Keep confidential records of all complaints in a separate file*

TIPS: ■ An impartial investigation cannot be done by the person who is being complained about, and should not be done by a person who is perceived as biased.

■ Investigation records should be kept confidential and not combined with personnel records where others not involved in the investigation could read them.

■ **Done wrong:** Starting an investigation long after receiving a complaint. While not all courts have defined "prompt," a month later isn't.

 ASSURANCE OF IMMEDIATE AND APPROPRIATE CORRECTIVE ACTION

○ *Includes discipline proportional to the seriousness of the offense, including, but not limited to, oral or written warning or reprimand, transfer or reassignment, demotion, reduction of wages, suspension, discharge, and training and monitoring of harasser*
○ *Must have measures to prevent harassment and ensure that it does not recur*

TIPS: ■ While the policy should list possible options for corrective action, it does not have to be all-inclusive.

■ Many policies simply state, "A violation of this policy may include consequences up to and including termination."

■ **Done wrong:** We have read policies that fail to mention any corrective action, and, therefore, fail to have any "teeth." To be effective, policies have to have a "bite" to them.

PROTECTION AGAINST RETALIATION

- ❍ *Provides assurance against retaliation for people bringing complaints or participating in investigations*
- ❍ *When investigating, each party should be reminded of no retaliation*
- ❍ *Should have a process for scrutinizing employment decisions during and after investigation*

TIPS:
- ■ Make sure to provide a reporting procedure for retaliation. It usually makes sense to have it the same as for reporting harassment.
- ■ Consider making another "zero tolerance" statement, but this time it should be about retaliation.
- ■ Have a process in place for scrutinizing employment decisions during and after a harassment investigation. Otherwise, you may find out that one of your managers has fired the complaining party within a month of the complaint with no documentation of poor performance . . . just after you have received an EEOC charge for retaliation.
- ■ **Done wrong:** Fail to have language on retaliation and reporting retaliation. Courts may find it reasonable for a plaintiff not to report harassing conduct if there was reasonable fear of retaliation.[5]

 LOGISTICS

- ❍ *Should establish, publicize, disseminate, redistribute, and enforce policies and procedures*
- ❍ *Letter from management should convey the seriousness (i.e., a message from the top officials)*
- ❍ *Policy must be understood by all employees*
- ❍ *Policy should be posted and in handbooks*
- ❍ *Provide training for managers, supervisors, employees, and investigators (how to properly interview witnesses and evaluate credibility)*
- ❍ *Evaluations of personnel should reflect compliance with harassment prevention policy*
- ❍ *Should have screening of supervisory promotional applicants to ensure they are not violating harassment prevention policy*
- ❍ *An acknowledgment form of the policy must be signed by all employees and retained by the employer*

TIPS:
- ■ Circulate the policy frequently so that every employee knows it exists and knows how to report harassing behavior.
- ■ Think about translating your policy into the language(s) your employees speak so that it will be understood.

- Conduct harassment prevention training based on the policy, not on facts and theories on harassment. Keep it practical so that participants know how to apply every section of the policy.
- **Done wrong:** Draft a policy, but fail to disseminate it. That's what the defendants did in *Faragher* and they ended up before the Supreme Court, as losers!

LESSONS LEARNED

To: Not Another Policy
From: The Trainers
Re: Lessons Learned

Dear Not Another Policy:

Once you have whipped your policy into good shape, there are a few major lessons we want you to learn. We are not subtle.

Distribute the policy. Distribute the policy. Distribute the policy. Distribute the policy. Distribute the policy. Distribute the policy. Distribute the policy. Distribute the policy. Distribute the policy. Distribute the policy. Distribute the policy. Distribute the policy. Distribute the policy. Distribute the policy. Distribute the policy. Distribute the policy. Distribute the policy. Distribute the policy. Distribute the policy. Distribute the policy.

Train on the policy. Train on the policy.

Take prompt effective action to follow up with all complaints. Take prompt effective action to follow up with all complaints. Take prompt effective action to follow up with all complaints. Take prompt effective action to follow up with all complaints. Take prompt effective action to follow up with all complaints. Take prompt effective action to follow up with all complaints. Take prompt effective action to follow up with all complaints. Take prompt effective action to follow up with all complaints. Take prompt effective action to follow up with all complaints. Take prompt effective action to follow up with all complaints.

Signed,

The Trainers

Notes

1. EEOC Enforcement Guidance: Vicarious Employer Liability for Unlawful Harassment by Supervisors (6/18/99), EEOC Compliance Manual (BNA) (EEOC Enforcement Guidance). *See* Appendix D for a complete copy of the EEOC Guidance.

2. *See, e.g.,* Haugerud v. Amery School Dist., 259 F.3d 678, 699–700 (7th Cir. 2001) (plaintiff's failure to follow the harassment policy was appropriate since the policy permitted complainants to file a charge of discrimination with the state agency in addition to or instead of using the internal complaint procedures).

3. Stay away from language that may create an employment contract. For example, "Our intention is to keep you in your position for a lifetime and so we have created a policy about respect." Also, stay away from language that is impossible to meet. For example, "You will not be retaliated against."

4. Heads up, though: Some states, Connecticut, California, and Massachusetts among them, have very specific requirements. For example, in Massachusetts, the Massachusetts Commission Against Discrimination has a model policy that suggests you use specific names and telephone numbers, and yearly dissemination of the list is required by statute. Mass. Gen. Laws Ann. Ch. 151B, § 1, 3A (2001). *See* Appendix F.

5. *See* Thomas v. BET Soundstage Restaurant, 104 F. Supp. 2d 558 (D. Md. 2000) (question of fact as to employee's reasonableness where she knew about reporting procedure of policy, but evidence suggested that her manager discouraged complaints and her supervisor was infamous for firing people); Katt v. City of New York, 151 F. Supp. 2d 313 (S.D.N.Y. 2001) (issue as to whether police officer feared retaliation for reporting harassment). *See also* Anderson v. Deluxe Homes of Pennsylvania, Inc., 131 F. Supp. 2d 637 (M.D. Pa. 2001).

3

One Canoe, No Paddle: Developing Content for Harassment Prevention Training

Dear Trainers:

OK, I finally got the folks upstairs to recognize that we could do some good by training our managers and supervisors and frontline employees. Wouldn't you know it, I got an emergency call today that they need harassment training at one of our facilities downstate next week! I think that there are some specific things that ought to be in the training but what are they? Should I read Title VII to them? How about managers and supervisors? Is there something different I should be telling them? Why? I need the basics. I feel like I am struggling up the river with . . .

Signed,

One Canoe, No Paddle

. . . .

Dear One Canoe No Paddle:

You are right to be concerned about the challenge that lies ahead of you. The topics and issues that should be covered in harassment training give your organization the greatest protection from harassment claims and legal liability. Using the Supreme Court cases as a guidepost, training must complement an employer's harassment prevention policy. This is the foundation of any potential affirmative defense. Now that you know a little bit about the legal issues involved in training, we are going to move on to what issues should be addressed in the training. We will share some basic ways to deliver the teaching points (more are in Chapter 4) and even look at questions participants commonly ask that will help you drive home your messages. Simply stated, start with your policy. Pull it out, dust it off, and read it carefully. Leave Title VII behind.

Signed,

The Trainers

Overview

This chapter will give you a main focus for your training—your harassment prevention policy. And based on your policy we are going to help you identify the basic content that should be in your harassment training. As we go along, we will give you an explanation of the "teaching points" for each topic, as well as some practical tips.

Let's talk about the focus of your training and whether or not you should read aloud the Civil Rights Act of 1964 (or Title VII) to your captive participants. Simply stated, "Don't!" Legal statutes such as Title VII, state anti-discrimination laws, and case law may seem helpful at first glance. However, the focus of harassment prevention training should be to communicate your organization's harassment prevention policy and its procedures for preventing, communicating, and reporting harassment. The goal is to build employee awareness and develop the skills needed to understand and avoid violations of your policy. This is what will help your organization avoid EEOC charges, litigation, and liability. Therefore, a law review course is not necessary. But to reach your goal, each part of your policy must be thoroughly reviewed in the training with an emphasis on understanding how to apply the policy in everyday situations. Legal statutes are definitely secondary, unnecessary, only marginally relevant, and, quite frankly, boring. Leave them behind.

Why focus on your policy instead of the law? First, the training is about teaching the organization's policy, which should be stricter than the law. An employee may violate the policy without necessarily breaking the law. This is good for the organization because it means that not every policy violation is an illegal act that runs the risk of exposure to compensatory and punitive damages. Second, what judges or juries may consider illegal behavior is not a good, practical guideline for employees to follow. There are too many different judges and juries taking into consideration unique facts and then drawing the legal conclusion of whether or not harassment occurred. When we train, we tell participants, "Don't go there."

What employer wants to spend hundreds of hours, $100,000-plus dollars in legal fees, an abundance of aggravation, and risk a six-figure verdict to find out a judge or jury's opinion about what is, or is not, harassment? It is just too risky to set the conduct bar as a moving target. The safer and more stable place to set the conduct bar is with your policy. So, focus the training on what the policy says and how it fits into the culture of your organization. Debating the merits of Title VII or state laws, or trying to interpret nuances of the latest court decisions and jury verdicts, only steers the training session away from your policy. Instead, give participants a firm grip on the conduct that is required of them in day-to-day work-related situations.

If the focus of harassment training is your policy, respect for one's co-workers should be the underlying theme. Harassment training should result in employees, managers, and supervisors realizing the need to treat others with respect so that the workplace is a professional, productive, and comfortable environment. This theme of mutual respect must be present at every aspect of the course and interpreted from every section of the policy. For when the trainer talks about the organization's policy, the trainer should not only be providing employer-mandated guidelines, but also providing participants with the necessary tools to make judgments about respectful behavior. The ultimate employer goal of mutual respect both on and off organization premises (in work-related situations) will be more easily achieved by coaching employees on taking responsibility for their own conduct and developing good judgment when interacting with co-workers.

But one word of caution about training on respect. We review a fair amount of harassment training—or, putting it more accurately, what some try to pass off as harassment training. Often this training is billed as harassment training but has the following fatal flaws:

- It doesn't provide specific clues as to what conduct might be considered harassment.
- It takes a very permissive approach to offensive comments (i.e., they are "innocent" or cultural misunderstandings that are not "intended" to be offensive).
- It doesn't teach much content, but rather a process for "working things out."
- It doesn't mention your harassment policy.

Well, you may ask, "Why are these fatal flaws? Isn't a process for dealing with problems of disrespect a *good* thing?" Yes, but not in place of your fundamental building block, your essential foundation piece—your policy against harassment!

Once, while we were doing a seminar for HR professionals, a woman came up to us afterward and told us that she had been doing harassment training on "respect" for a newly acquired unit. She said she spent over two hours of the training trying to persuade many of the participants that it isn't nice to be disrespectful toward people in the workplace based on their sexual orientation. She said, "I realized after you talked about harassment training that I could have accomplished more in two minutes than I did in two hours by telling them that being disrespectful to people based on their sexual orientation violates our harassment policy and can subject you to many unpleasant consequences." We agree!

Not having a focus on specific employment-related laws does not mean you are not interested in limiting legal liability or in being legally accurate. Make no mistake about it—you are. For most employers, training employees in order to reduce potential liability is certainly one part of the goal. After all, you want to be able to use your affirmative defense in those situations, hopefully few and far between, where the errant manager, supervisor, or co-worker may have missed or ignored the message. This of course makes good business sense. But since *Faragher/Ellerth* and the affirmative defense, your training, in addition to accomplishing all of your training goals, must also accomplish legal goals. One of those goals is that the content of your training, while not legally focused, must be accurate and complete from a legal perspective. You do not want to have your training challenged later on the grounds that the content that should have been there wasn't, or that the content that *was* there was inaccurate.[1]

A clear definition of the type of conduct prohibited by your harassment prevention policy is essential to ensure that employees understand exactly what they should and should not do. But where to begin? We have reduced the EEOC's elements from the policy chapter (Chapter 2) to four major areas that we need to cover in training:

1. Identifying Prohibited Conduct (Quid Pro Quo and Hostile Work Environment)
2. The Reporting Process
3. Retaliation
4. Special Obligations of Managers and Supervisors

Additionally, if you are in an union environment, or in the public sector, there are special issues that you may need to consider. At the end of this chapter we will address the special considerations of those workplaces.

What You Need to Know . . .

Identifying Prohibited Conduct

Prohibited conduct is the heart of the harassment prevention training. Training that solely provides a definition of harassment and declares that harassment is prohibited is too "bare-bones." Training sessions must be used to further explain the policy. We have worked with organizations that have "trained" employees by saying, "Just don't do anything disrespectful to anyone." Later, they are often surprised to find that "respect" is an abstract idea that does not have one definition. A clear definition of the type of conduct prohibited by your harassment prevention policy is essential to ensure that employees understand exactly what they should and should not do. But, you ask, the policy cannot describe every kind of conduct under every circumstance that is prohibited, can it? Of course not, but that is where training comes in.

Training should "flesh out" the black, the white, and the gray areas of conduct. For example, what is absolutely acceptable (the white), what is absolutely unacceptable (the black), and, more common for most workplaces, what is the conduct that resides in some middle ground? It is that middle area, sometimes called "the gray area," that requires personal judgment about conduct that is risky under your policy, perhaps resulting in a policy violation. For example, it may be perfectly acceptable for an employee to ask a co-worker out on a date.[2] But if the co-worker declines and the employee asks again, is there a risk of violating the policy? What if the co-worker declines again and the employee asks for the third time? Is it even riskier? Obviously, these types of examples for every situation cannot be written in your policy. But this is one of the functions of training, to explore acceptable and prohibited conduct and where you want employees to "draw the line."

Remember too that when we are talking about prohibited harassment we are not only talking about conduct of a sexual nature—policies and training must also deal with all forms of harassment prohibited by federal, state, and local laws. For example, the policy and training must also address harassment based on race, color, religion, national origin, age, and disability. Nonetheless, sexual harassment is sufficiently different and perhaps sufficiently severe to warrant special attention. The most egregious form of harassment, and one that is unique to sexual harassment, is called "quid pro quo."

Quid Pro Quo

 TEACHING POINTS

- ▶ **Definition of Quid Pro Quo Harassment**
- ▶ **Sexual Conduct by Managers or Supervisors**
- ▶ **Not Just About "Sex"**
- ▶ **Unwelcome Threat to Job or Job-Related Benefits**
- ▶ **For Managers and Supervisors: Subtle Quid Pro Quo Issues and Dating Relationships**

DEFINITION

One form of harassment, specifically used to describe a form of sexual harassment, is "quid pro quo," a Latin term meaning "something for something." Quid pro quo harassment occurs when an individual's submission to or rejection of sexual advances or conduct of a sexual nature is used as the basis for employment decisions affecting the individual, or when the individual's submission to such conduct is made a term or condition of employment. Quid pro quo harassment is often known as the "put out or get out" bargain, the classic sexual harassment situation where a manager/supervisor demands sex in exchange for continued employment.

For the most part, explaining quid pro quo sexual harassment to employees is not a difficult undertaking. The concept of demanding sex for a raise or sex to keep your job is a simple one to put one's mind around. It is easily understood as wrong and in violation of the policy and even the law. To put it differently, very few participants faced with a quid pro quo situation are going to think the behavior is or should be permitted by your policy. Thus, spending a lengthy amount of time on this concept, particularly with employees, is usually not necessary. When talking about quid pro quo harassment with employees, we especially focus on the many avenues of reporting available under the policy. For example, in the situation in which someone is being harassed by his or her manager or supervisor, we train that he or she does not have to go to that manager or supervisor to make a complaint. Rather, the person being harassed should go to someone else listed in the reporting procedure. Remember, too, how closely tied these topics are to the issue of retaliation. While we cover retaliation a little later on in our training, we are always prepared, and you should be too, to talk about the policy's prohibition against retaliation for reporting inappropriate conduct.

SEXUAL CONDUCT BY MANAGERS OR SUPERVISORS

Only management or supervisory employees, i.e., someone who can make or bring about tangible employment actions such as firing, demoting, blocking promotions, transferring, or providing performance evaluations, can commit this kind of unlawful sexual harassment. To put it differently, anyone in the organization with the power and authority to affect the victim's terms and conditions of employment can potentially create quid pro quo harassment.

TIPS: ■ This is why managers and supervisors must be specifically trained on the grave consequences and tremendous exposure to liability that such conduct can create.

NOT JUST ABOUT "SEX"

Quid pro quo harassment is not just about sex, but can occur whenever sexual favors are bargained for. Thus, requesting a back rub, a date, or a kiss also can lead to quid pro quo harassment claims. While quid pro quo harassment usually involves a bargain for sexual favors that results in negative actions, it may also result in rewards, such as raises, promotions, job-training opportunities, favorable performance reviews, benefits, or other perks. In other words, quid pro quo does not always mean "sleep with me or you are fired." It could be "sleep with me and you'll get promoted."[3]

TIPS: ■ When we train about this element we remind participants that the definition of sex is broad. We often are asked to explain what that means and we reply, "It is not just about what we commonly know as having sex, but also anything that may be considered to be of a sexual nature." If participants require further clarification, we simplify it further, "Anything sexual." Enough said-usually everyone understands that concept.

UNWELCOME THREAT TO JOB OR JOB-RELATED BENEFITS

Legally, conduct must be perceived by the victim as "unwelcome" to constitute quid pro quo harassment. This naturally leads us to walk around asking, "How can I tell if something I said or did is considered legally unwelcome?" The practical answer is, "You really can't." While we briefly discuss "unwelcomeness" in the hostile work environment section, we focus very little on it in training. In many cases, it's just too hard to determine whether something was or wasn't perceived as welcomed. You are safer to focus on conduct prohibited by your policy.

TIPS: ▪ Again, it is too risky to have employees use the legal standard of "welcomeness" to guide their conduct. Therefore, in the training we warn participants, "Don't go there. Keep your conduct consistent with the policy if you want to avoid consequences."

FOR MANAGERS AND SUPERVISORS: SUBTLE QUID PRO QUO
ISSUES AND DATING RELATIONSHIPS

In training managers and supervisors, tackling quid pro quo and subtle issues such as dating of subordinates can be handled in a number of ways. The first question that must be addressed is whether the organization has a policy forbidding consensual relationships between employees and their managers or supervisors. Under the assumption it does not, you should present dating a subordinate as a risky endeavor that can raise potential subtle quid pro quo issues as well as other problems that can lead to policy violations. Explain how dating a subordinate can be risky because if the relationship breaks up the subordinate may claim to have been pressured to enter the relationship and/or claim it was never consensual.

While we usually suggest you stay away from discussing cases to make legal points, when it comes to giving "teeth" to the message about quid pro quo harassment, real-life stories and court cases may be the best weapon to hammer home the point.[4] Of course, if the organization has a separate nonfraternization policy that prohibits such relationships, you need to clearly say that such relationships are prohibited, and distribute that policy. You may also want to have participants sign off on a policy acknowledgment form that they received the policy. Then move on.

 TIPS: ▪ When discussing real stories, it is always a best practice not to use any names or identifying information and, of course, never from your own organization. Even though reported cases are public information, we have a practice of not using the name of the organization, but rather the name of the industry (e.g., "a major retailer," "an auto parts chain store," etc.). You never know who is in the audience and their connection to the story or case.

Hostile Work Environment

 TEACHING POINTS

- ▶ Definition of Hostile Work Environment
- ▶ Other Protected Categories
- ▶ Whom Does the Policy Cover?
- ▶ Welcome or Unwelcome Conduct—Who Decides?
- ▶ Severe or Pervasive

▶ **Not All Unwelcome Conduct Is Harassment**
▶ **Off-Premises Conduct**
▶ **"Bystander" Harassment**
▶ **Gender-based Harassment**
▶ **Conduct That May Create a Hostile Work Environment**

DEFINITION

We use the following definition: Hostile work environment harassment is any conduct that has the purpose or effect of unreasonably interfering with an individual's work performance or creating an intimidating, hostile, or offensive working environment. As mentioned before, whether a hostile environment legally exists depends on a number of factors, including the severity and frequency of the inappropriate conduct, and whether the conduct was both subjectively and objectively unwelcome.

Unlike quid pro quo, which only someone in authority can create, a hostile work environment can be created by anyone in the workplace. A hostile environment can result from any unwelcome conduct of co-workers, managers, supervisors, executives, customers, vendors, or anyone else with whom the victimized employee interacts on the job. Hostile work environment harassment can also occur when threats of adverse job consequences (such as those for quid pro quo) are made, but are not carried out. In addition, individuals or groups who interact in a sexual manner or whose behavior is unwelcome may create a hostile environment for another employee who is merely a bystander. In other words, an employee may be the victim of a hostile work environment if he or she merely overhears inappropriate comments.

In our experience, talking about a hostile work environment generates the most discussion and takes the most time in a harassment training session. Generally, participants understand the issues involved in explicit quid pro quo harassment, but hostile work environment has some unclear boundaries that need to be reviewed and answered. And, of course, this is the most common type of harassment that is found in the workplace, making it a very important part of the training.

 TIPS: ▮   "Is it okay to hug someone at work when I see them or will that create a hostile work environment?" We think that the definition of a hostile work environment is useful for giving the big picture, but really, what everyone wants to know is: Is "this" okay? Is "that" okay? Your training can't just define hostile work environment, it must use the definition as the "bones" and then put the meat on the bones. Beginning with the definition, you can then extract from it the subsequent teaching points. We sometimes have our definition of hostile work environment harassment prewritten on an easel pad and the key words written or underlined in red. Then we ask the participants to answer the "Is this okay?," "Is that okay?" questions using the collective wisdom of the group and the techniques outlined under "Conduct That May Create a Hostile Work Environment" in this chapter (page 47).

OTHER PROTECTED CATEGORIES

It is important that participants are aware that the anti-harassment policy not only covers sexual harassment, but harassment based on all protected characteristics. While *Faragher* and *Ellerth* dealt with issues of sexual harassment, courts and the EEOC have reasoned that the affirmative defense applies to harassment by managers and supervisors based on race, color, religion, sex (whether or not sexual in nature), national origin, age, or disability.[5] Many states and policies also cover sexual orientation, marital status, and other characteristics (medical conditions and political affiliations, to name a few).[6] Thus, employers should establish harassment prevention policies, complaint procedures, and training covering all applicable forms of unlawful harassment. While discussing sexual harassment issues often is the emphasis of a training program, the program also must stress that the employer's policy covers harassment based on other protected characteristics.

 TIPS: ■ Jokes. Perhaps the most effective way to illustrate harassment based on characteristics other than sex is to focus on common workplace conduct such as making jokes. We have all heard jokes that can be perceived as offensive based on any and all of the protected characteristics. Such jokes are examples of what violates policies. If a joke, card, or gag is based on a protected category . . . just don't go there.

In illustrating this, we generally use "humorous" birthday cards about age to make our point.[7] The usual custom of sending around birthday cards with obnoxious comments about how old the person has become is inappropriate and has led to harassment complaints.[8] While everyone has heard or told off-color jokes about virtually every protected group, age-related cards don't, at first glance, seem to have the same inappropriate stigma as jokes about race or ethnicity, for example. That is why we use age—to help raise awareness of the harassment possibilities inherent in age jokes, and to make the point that what seems inoffensive to some, may be offensive to others.

■ In discussing jokes in training, never tell a joke to its punch line. The last thing you need is a participant repeating the whole joke about "the rabbi, the priest, and the sexy stripper," having heard it in your training. Stop the joke after the setup and just use it as an example of where "not to go."

▪ "Are we in diversity training or harassment training?" Talking about people with many different protected characteristics is a great opportunity to connect to any diversity training that your organization does. Based on our diversity we all bring something unique, useful, and important to the organization. And it is important that participants understand that harassment is not just a "man-woman" thing. We find that participants are less resistant if they understand that this is really about all of us—that all employees are protected from harassment, whether because of race, color, religion, sex, national origin, disability, or age.

WHOM DOES THE POLICY COVER?

Who can be a victim of harassment? Who can be a harasser? Not only must the training discuss what is prohibited conduct, but it must also address who is protected from such prohibited conduct as well as who can perpetrate the conduct. Emphasize that the policy applies to both males and females. And reiterate that males can just as easily be the target of harassment, including sexual harassment. A recent survey indicates that approximately 87 percent of all sexual harassment claims are filed by women; 13 percent of all claims are filed by men.[9] We sometimes use these numbers to show that men are not always the harassers and to help participants, particularly managers and supervisors, understand that they must take prompt, effective action on *all* harassment situations, even if the alleged harasser is a woman.

In addition, while many understand that sexual harassment can occur between persons of the opposite sex, sexual harassment can also occur involving persons of the same sex. Accordingly, managers and supervisors must be able to recognize same-sex harassment.

Finally, it is important to stress that the policy not only covers employees, but anyone that steps into the workplace or conducts business with the employer, from vendors to clients to delivery people.

 TIPS: If there is one mistake we see over and over in training, it is the failure to mention clients, vendors, independent contractors, and other third parties. Depending on your business, your employees may come in contact with any or all of these people—and, because third parties are not employees, may mistakenly think:

▪ that the third parties can be harassed; or
▪ that if they are being harassed by a third party, they have no recourse.

Of course, neither is true. Make sure to point this out in your training.

WELCOME OR UNWELCOME CONDUCT—WHO DECIDES?

Here is how "unwelcomeness" arises in hostile work environment claims: what is acceptable, amusing, or welcome behavior to some people may be offensive or unwelcome to others. Legally, harassing conduct must meet two requirements. First, it must be subjectively perceived as abusive by the person(s) affected. In other words, the victim must *personally* have found the conduct offensive. Second, it must be objectively severe or pervasive so that a "reasonable person" in the shoes of the victim would find it hostile or abusive.[10] This means that an objective standard, what the "reasonable person" might think, also will be applied. This should lead into a discussion of what is "reasonably offensive." For example, a reasonable person is not overly sensitive or overly insensitive, but rather represents how the average person would experience the conduct.

We do not spend an inordinate amount of time on this problematic area because we believe the focus of the training should not be on whether the affected individual was reasonable in being offended by the conduct. This "reasonableness" standard is often hard to gauge and unknown at the time of the conduct. Rather, the focus of the training should be on avoiding any behavior that could reasonably be perceived as offensive.

TIPS:
■ Emphasize that just because an employee does not immediately state that she/he is bothered by conduct does not mean that the conduct is welcome. Likewise, laughter does not mean that the conduct is welcome, since co-workers often laugh at an offensive joke because they are uncomfortable, or just to get along.

■ Courts assess behavior through the eyes of the victim and not through the intentions of the harasser. Therefore, employees need to become more sensitive to the perceptions and feelings of others in order to avoid unwelcome conduct, policy violations, and possible litigation. The issue to emphasize is that unwelcome conduct is viewed, not by what the speaker intended, but how someone was affected by that person's conduct.

■ To visually illustrate that people see things from different perspectives depending on whose shoes they are standing in, we sometimes use a male and a female shoe when debriefing role plays, case studies, or video vignettes. (See Chapter 5 for training tools.)

■ You may also want to address the difference between quid pro quo and hostile environment sexual harassment—or actions that affect job benefits as opposed to actions that create an intimidating, hostile, or offensive working environment. While these terms are not as important in a legal sense as they once were, they are still useful terms to describe the different types of impermissible conduct.

SEVERE OR PERVASIVE

For hostile environment harassment to violate Title VII, the conduct must be "sufficiently severe or pervasive to alter the conditions of employment and create an abusive working environment."[11] This means that the conduct must be so offensive and/or persistent that it affects the employee's job and work environment. A judge or jury will often consider the following factors in deciding whether behavior is severe or pervasive enough to create a hostile work environment:

- the frequency of the unwelcome discriminatory conduct;
- the severity of the conduct;
- whether the conduct unreasonably interfered with work performance;
- the effect upon the employee's psychological well-being;
- whether the conduct was physically threatening or humiliating; and
- whether the harasser was in a position of authority to tangibly affect a person's job—such as by hiring, firing, or demoting.

"How can I tell if my conduct is severe or pervasive enough to violate the law?" We recommend that you touch only very briefly on the issue of "severe and pervasive." First of all, this is a legal issue used to determine whether a particular set of facts constitutes a successful harassment claim and legal liability. It does not determine whether a violation of your policy has taken place, because conduct can violate your policy without rising to the higher level of being severe or pervasive.

TIPS:
- It is not useful to try to definitively define "severe and pervasive" in the training. For example, what exactly is "severe"? In determining whether something is "severe," the "severity of the conduct" is considered. Well, now, that is just *not* useful to your participants. And how many incidents equal "pervasive"? One "severe" plus two minor incidents? Not to mention, who's counting?
- All that participants really need to know is that even one incident of inappropriate conduct that violates your policy can lead to consequences. Whether an employee's action(s) violate the law is irrelevant from the training perspective when the goal is to keep them from violating the policy, which is stricter than the law. For everyone, the

motto really is, "Don't go there." We like to tell participants that this is good news. You do not have to determine whether the conduct in question meets the legal definition of harassment. Rather, all you have to do is stay away from any conduct that would violate the policy and report any conduct that does. If you "don't go there" at the beginning, you won't end up in a courtroom at the end.

■ "I did not intend for my conduct to be offensive. Doesn't that matter?" Does intent matter? The short answer is "no." The long answer is "absolutely not." But many of your participants will be surprised by such news. Since we are looking at things from the shoes of the victim, it doesn't matter what the alleged harasser intended. We tell participants that neither the law nor our policy recognizes the harasser's intent to be funny or just horse around to have a "good time" at work. The focus is on how the victim perceives the harasser. That's how these situations and cases are reviewed. Just don't go there.

NOT ALL UNWELCOME CONDUCT IS SEXUAL HARASSMENT

We have always found that is it important to explain what is *not* sexual harassment. For example, conversation between employees and managers or supervisors on work rules and performance management, while potentially unwelcome to employees, is not necessarily unwelcome conduct based on race, color, religion, sex, national origin, age, or disability. Unless, of course, it involves favoritism or demeaning comments based on protected characteristics, for example, "only Asian women have to follow the rules," or "you Asian women do not perform well because you do not follow the rules." The key component to address is that sexual harassment must be harassment based upon a person's sex, just as racial harassment is based on a person's race. Behavior that is unwelcome for other reasons, like feedback about his or her job performance that an employee does not want to hear, is not prohibited by the harassment policy.

 TIPS: ■ "It seems like the best thing to do is shut up and not speak. Right?" Participants in harassment training often opine that "it isn't possible to talk to anyone anymore." It is good to debunk that myth and explain that exchanging pleasantries, engaging in discussions on obviously inoffensive and nonsexual topics, and participating in normal work relations do not constitute harassment. It is still okay to talk about books, movies, sports, weather, current events. Let's be reasonable and . . . go there.

OFF-PREMISES CONDUCT

One teaching point that is often overlooked when discussing a hostile work environment is the meaning of the term "work environment."

From a legal perspective, inappropriate work-related conduct may be part of a hostile work environment claim no matter where it occurs. Off-site work-related activities, often not considered by employees and employers to be part of "work," are a prime breeding ground for inappropriate behavior. Many employees view their behavior as susceptible to employer "monitoring" only inside the four walls of the workplace structure. And many management personnel feel and/or assume that they have no responsibility for, or authority over, employees outside the workplace. Thus, you should spend time on this concept with managers and supervisors as well as employees.

Participants need to understand that the policy applies not only in the work setting (offices, plants, warehouses, etc.) but also in some locations that are outside of work, but work-related. Indeed, off-premises conduct with some connection to the workplace is common in legal claims. Whether in a bar, at a company outing, or in the parking lot, after work hours or on a business trip, conduct that has an adverse effect on the working environment may violate the policy.[12]

 TIPS:

■ "When I've clocked out of work I'm on my own time. You don't own me, right?" Off-premises conduct is a common problem in some workplaces. But how far do the tentacles of an anti-harassment policy really reach? This is a difficult question and one frequently asked and argued by participants. It is also a difficult message to deliver because there no easy answers. For example, what if a manager or supervisor observes an employee violating the policy at a non-work-related function on a weekend night? The policy may not apply. Or what about the ever-famous holiday party or sales conference paid for and sponsored by the organization? The policy may apply. In the middle we have incidents that happen outside of work, yet affect work—for example, telling the dirty joke from the bar on Saturday night around the water cooler on Monday morning. And incidents that occur at the bar on Thursday night after work where the whole department gathers every week as arranged by the organization's e-mail system. Does the policy apply?

■ There may not be a clear-cut answer to what is off-premises conduct, but in trying to develop good judgment about what might offend a reasonable person, we need to help participants understand the issues that affect the analysis. Courts will look at certain elements to determine if the "off-premises conduct" was work-related:

– Did the organization pay for or sponsor the event?
– Is it a regular gathering of employees?
– Is it organized using organization resources (like the e-mail system)?
– Are there a number of employees there? We sometimes use this rule of thumb: "If you look around and everyone you see is a co-worker, the policy probably applies—no matter where you are."

- Does the event get talked about a lot back in the workplace?
- Was work discussed during the gathering?

"BYSTANDER" HARASSMENT

Participants in harassment training are often surprised to learn that even if the person they are talking to welcomes the dirty joke or the racist e-mail, a third person, a "bystander" who is not the direct recipient but overhears the joke or finds the e-mail on the printer, may also make a complaint under the policy. Remember, a hostile work environment really is about the "environment" and everyone in it.

TIPS:
- Sometimes we talk about an innocent bystander getting hit by "harassment shrapnel," something not directed at them, but that they find offensive.
- "So how can you tell a dirty joke and have fun at work anymore?" Once we had a participant ask, "What if I am telling a joke and I invite my friend into my office so no one else can hear and I ask him if it's okay to tell a dirty joke and he says yes." At which point a person in the back piped up, "If you have to ask, don't tell." It's a really good answer. Remember, there are still ways to have a good time at work without violating the policy. Coffee breaks can be filled with chatter about sports, movies, the weather, books, and of course . . . work. (Don't be afraid to remind employees what they are supposed to do at work. Work.)
- Participants often have difficulty believing that someone who is not involved in a conversation has the right to make a complaint just because they overheard something offensive. When we get resistance on this point, we take a minute to give participants this reminder:

  H – Hostile; harassment often feels hostile in the perception of the victim;

  W – Work; it occurs at work or work-related events;

  E – Environment; it's about the environment and everyone in it, not just what happens between two or three people.

GENDER-BASED HARASSMENT

Another form of hostile work environment harassment that should be covered in training is gender-based harassment. It is harassment that does not necessarily involve sexually oriented language or conduct, but does involve other offensive, gender-based language or conduct directed at someone simply because she (or he) happens to be a woman (or a man). It could include comments such as, "Women shouldn't be taking jobs away from men," or "You men think you run this place, but it would fall apart without us women," or "You women should be making the coffee and cleaning the bathrooms." It could also include giving specific assignments based on gender. The key point for participants to understand is that harassment based on gender does not have to be motivated by sexual desire.

CONDUCT THAT MAY CREATE A HOSTILE WORK ENVIRONMENT

This is the heart of the harassment training and it should help participants assess their own conduct and develop the judgment necessary not to "cross the line" from appropriate into inappropriate conduct. Knowing when one crosses the line can be difficult, mainly because while there are a few "black and white" issues that are easily identified ("Sleep with me or you're fired," or racist jokes,[13] or calling a Hispanic person "Taco"), there are hundreds of situations that are more ambiguous.

How to begin? As always, have participants assess behavior through the eyes of the *recipient* of the behavior, not through the eyes or the intentions of the alleged harasser. As stated before, we need not get bogged down with the elusive "reasonable person" standard, which is determined at trial by the judge or jury. Rather, we train on the broader point that we should view behavior from the victim's point of view and not the harasser's. Thus, conduct that many people view as "harmless fun" may violate the policy because others find it offensive.

Once we know that we have to look at the conduct from the perspective of the recipient, we need to get a general idea of what is and is not appropriate conduct under our policy. Of course, lists in policy statements can give examples of what is not permitted; but they often do not show what is permitted. They also don't give participants the tools to determine for themselves whether behavior will or will not violate the policy. This is critical because not every example of inappropriate behavior could possibly be covered in any list without making the policy hundreds of pages long.

For example, most policies prohibit inappropriate touching. But what exactly is that? We all probably know what body parts that absolutely includes, but what about shaking hands? Is that inappropriate? Or patting someone on the back and saying, "Good job!" Both are okay, but you wouldn't know that by simply reading the policy. You have to put flesh on the bones of your policy through training or it is meaningless as a day-to-day guide to conduct.

Of course, the ultimate goal is to give participants a better framework for determining whether conduct is appropriate in the workplace, while also making the explicit point that risky conduct that might violate the policy should also be avoided. Certainly, teaching judgment is no easy task. And there are no easy answers. We use all the tools at our disposal to help participants create a framework for determining appropriate workplace conduct. But regardless of the tool, in our experience context is the key component to learning judgment. By context we mean that participants must be able to wrap their mind around the whole spectrum of behavior—from good behavior to bad behavior to everything in between. They must know that there is a spectrum of behavior. Participants can't see the middle (the everything in between) without first understanding and mastering the ends. So, regardless of the tool, we are always careful to first give participants examples of conduct that fall on either side of the spectrum, and then we attack the problematic middle.

The following tips will give you a better understanding of what we mean.

 TIPS: ▪ There are numerous tools that are very helpful in training on hostile work environment. The tools we like the best are those tools that force participants to work together to thoroughly discuss specific and realistic examples of workplace behavior. One such tool we call the "danger zones." Danger zones are dangerous areas where you must use caution. The following list of danger zones is a great place to start and will easily lead into other harassment issues, such as bystander harassment and welcomeness:

- comments on personal appearance
- work-related, off-premises conduct
- nicknames
- stereotypes
- touching
- dating/initiating personal relationships
- retaliating
- cartoons, posters, pictures, apparel, T-shirts
- e-mail/Internet
- jokes, sarcasm, or innuendo about race, color, religion, sex, national origin, age, disability, sexual orientation, or any other protected category

▪ Another tool to help give participants a better framework for determining whether conduct is appropriate is the concept of a "conduct spectrum" or a line along which conduct falls. For example: Draw a straight line with "OK" written on the far left side and "Not OK" written on the far right. Explain that conduct that falls on the left is behavior that is "OK" under our policy and conduct that falls on the far right is conduct that is "Not OK."

Write out several separate short hypothetical situations based on the danger zones or any other conduct that you want to address.

For example, you could write, "Chris says to Lee, 'I love how you look in those jeans!'" Or, "Pat says to Chris, 'I love your tie!'" Or, "Jackie says to Lee, 'I love how your behind looks in those jeans!'" What is appropriate, inappropriate, and in the "gray" area (the middle of the line or conduct spectrum)?

■ This kind of line or conduct spectrum is an important tool to illustrate that not all situations are easily identified as either appropriate or inappropriate; that not all situations have a right or wrong answer, and, in fact, may be perceived in different ways by different people. It is the conduct that falls in the middle of the line that proves to be the most valuable teaching device. These situations may not always violate the policy, but certainly, given the right set of circumstances, can lead to risk. These are the behaviors that will force discussion among the participants. You can also use these hypothetical situations to discuss other teaching points such as intent and perception, by asking, "What if Chris did not intend to offend Lee? Does it matter?" (No, intentions don't count.) Or, "If Lee perceives the conduct as offensive, does it matter?" (Perceptions do count.)

■ Another tool that is effective in teaching participants what may or may not violate the policy is what we call the "family rule." Simply put, if you would not say or do something in front of a family member (daughter, mother, father-in-law), do not say or do it in front of someone you work with.

The Reporting Process

The reporting process is a key element of your training from both a legal and a training perspective. Participants should be told that when they feel that they have been subjected to behavior that may be in violation of the policy or have witnessed other employees being subjected to such behavior, they should report their observations to the human resources department, or their manager or supervisor, or whoever is listed in the reporting procedure of the policy. Explain that after complaints are made, the employer will investigate and take any necessary corrective action. Again, emphasize that no person will be retaliated against for filing a claim and that all complaints will be investigated thoroughly and promptly and will be handled as confidentially as possible.

TEACHING POINTS

▶ **The Organization Complaint Process**
▶ **Confidentiality**
▶ **Consequences**

THE ORGANIZATION COMPLAINT PROCESS

Simply posting rules forbidding harassment is not enough to stop harassment. Everyone must be proactive by reporting harassment and making every effort to stop harassing behavior. Remember that legally, the most important teaching point to address is the how, where, and who of reporting inappropriate conduct. An effective or ineffective reporting procedure will be the focal point of any court's analysis of the liability of an employer. With that much legal importance, it is imperative that the training not make short shrift of this topic, an easy thing to do. Many trainers and programs simply put up telephone numbers and/or names of people to contact in case an employee observes or experiences harassing behavior. This is not sufficient.

It is critically important that everyone know where to go to report a complaint. It is common that plaintiffs who bring harassment lawsuits have not reported their concerns to their employer. Under the affirmative defense, a plaintiff must show that the failure to use the employer's reporting procedures was "reasonable." Once you have trained, and once you have done a good job on this topic (and if you can prove it by documenting curriculum, attendance, participation, and distribution of your policy), you should be in a good position to defeat any subsequent claims that failing to report was "reasonable."

Thus, the complaint process should be reviewed in complete detail, exactly as it is written in the policy.[14] Also note that the courts may hold you to exactly what you have written as a reporting process, so make sure you are completely clear on what it is. Because of the *Faragher/Ellerth* decisions, employers now sometimes make their complaint procedures mandatory, i.e., they make it a requirement that an employee report any conduct that he or she feels may be in violation of the employer's policy. Such a requirement will make it less reasonable for an employee to fail to use an employer's complaint procedure and, accordingly, will make employers better able to use the affirmative defense to avoid liability. Similarly, if your policy states that complaints should be reported to managers and/or supervisors, make sure that employees know who they are. If you do not want "team leaders" or "leads" responsible for taking harassment complaints, make sure you make that clear in the policy and training as well.

It is crucial that the trainer specifically spell out where in the policy the reporting procedures can be found. We ask participants to find this section of the policy and highlight it—that way we *know* they have seen it. Tell participants that they have choices on who to report to. Finally, trainers should articulate that people and positions listed in the policy may also be used as a resource for questions regarding the anti-harassment policy and are not solely there for reporting inappropriate behavior.

Basically, anything an employer can do to familiarize employees with the complaint process will forestall later claims that it was reasonable to forgo the process because it was so complicated, foreign, etc. At the end of the training session there can be no doubt whatsoever that every participant is aware of where to turn to report inappropriate behavior.

TIPS:
- "I know what the policy says about reporting but whom should I really go to?" We often use an exercise in which we ask participants to read the reporting section of the policy and then we ask, "Does everyone know at least one person that you could go to if you wanted to report an issue of harassment?" Then we ask that on the count of three, everyone call out the name of that person. Then we ask, "Did everyone have a name that they could call out?" If not, make sure to follow up with those participants. This is very high-energy and leads to a lot of laughing, but at the end of the day, we know that everyone knows someone by name to report to—a very important fact if you ever have to take your training to trial for the affirmative defense.
- "How can I tell someone who is offending me to stop?" It is worthwhile to spend some time on how to respond to harassing behavior if you experience it. Trainers should emphasize the proactive aspect of the reporting procedures. The teaching point is that employees need to tell the harasser to stop or inform the harasser that his/her behavior is inappropriate *only* if they feel comfortable confronting the person. Responding to a harasser is not mandated by the policy (or the law), but merely a suggestion. An employee who does not feel comfortable telling the harasser to stop could simply report the behavior, pursuant to the terms of the policy.
- Think about using a "script"—some simple, nonpersonal way that one person can tell another to stop. One is, "Don't go there." Another is, "I think that crosses the line and might violate our policy." Many people don't do anything to stop harassment because they don't know what to say. Giving them a simple phrase or two to work with is a great tool.
- "What should I do if someone tells me to stop?" On the flip side, time should be spent on how the alleged harasser must act when an employee tells him/her that his/her conduct or language is unwelcome. We tell them to consider it a "gift" because such feedback will help them keep their job. Often there are situations in which co-workers offend one another, then claim they intended no offense and were only joking or kidding around. The point to get across is that

employees must respect the victim's request for the behavior to stop, regardless of whether the harasser feels the employee is being overly sensitive or does not understand that nothing offensive was truly meant by the words and/or the conduct. It is not a time to debate whether the conduct was truly inappropriate or not. Instead, the alleged harasser should stop the conduct, apologize, and never do it again, to anyone. Again, giving a simple script, something like, "I'm sorry. Thanks for letting me know. I won't do that again," can be very useful to all.

■ "Am I *required* to report harassment that I know about?" Managers and supervisors definitely are, but what about frontline employees? Look to your policy. If it says "you must" report, then you will use that exact language in the training. Be prepared for the questions, "What if I don't? What is the consequence?" We see very few employers who are actually disciplining employees who fail to report either their own or someone else's complaint, so you may have to finesse this question. We often say that you are strongly encouraged to do so— after all, a hostile work environment affects all of us and we cannot make our workplace a place of respect for everyone until we are all willing to stand up and "be counted." The policy can work only if each person plays his or her part; it is the responsibility of all of us to make the policy work.

CONFIDENTIALITY

Finally, the subject of confidentiality must be broached. Complete confidentiality is an impossibility and must never be promised, because complaints must be followed up with a prompt investigation. Participants need to know that they cannot ask for complete confidentiality. Managers and supervisors need to know they cannot give it. Instead, employers should explain that it will keep all matters confidential to the extent possible and that it expects its employees to similarly respect the confidential nature of complaints and investigations. Often participants will be taken aback to learn that an employer cannot and will not guarantee confidentiality. Thus, it is important to briefly explain, in general, how an investigation will be conducted to illustrate why complete confidentiality is impossible.[15]

TIPS: ■ This is a good opportunity to go through a hypothetical harassment complaint. This hypothetical should illustrate how an investigator will handle a complaint—from what types of questions the investigator will ask, to what documents the investigator will look at, to whom the investigator will want to talk to, to how the results/disciplinary action will be presented. Through this scenario, it will become apparent why confidentiality cannot be complete.

CONSEQUENCES

It is extremely important that participants understand that there are consequences for violating the harassment policy. Our sense of fair play generally requires that we let employees, managers, and supervisors know in advance if there are rules that have consequences and if those rules have changed. In some workplaces, the rules have changed. In others, the rules are the same, but not everyone is aware of them. Training is the perfect opportunity to let everyone know that there are rules, and that the violation of those rules can lead to discipline, up to and including termination.

 TIPS: ■ "Will I automatically get fired if I violate the policy?" Another point you may want to clarify is the meaning of "appropriate" action after an investigation of a harassment complaint. Does this mean termination in every case? No. Does it mean that the alleged victim gets to decide what the discipline is? No. The victim's sentiment that "I want her fired" is no more effective than "I don't want anything to happen to her." In both cases, the organization must make a decision about what is "appropriate" based on the facts of each case.

Retaliation

 TEACHING POINTS

> ▶ **What Is Retaliation?**
> ▶ **Retaliation Is Prohibited**
> ▶ **Process for Reporting Retaliation**

WHAT IS RETALIATION?

The concept of retaliation is a simple one. Retaliation is getting back at someone because he or she does something you don't like. In the harassment context, retaliation consists of any negative conduct in response to an employee's complaint of harassment or participation in a harassment investigation. Retaliation can include adverse employment actions, such as demotions, transfers, or firings, or more subtle types of actions like being subjected to tougher performance standards than other employees. Co-workers also can engage in retaliation through such devices as ostracism, teasing, threats, withholding information, unacceptable behavior, the silent treatment, or physical actions.

RETALIATION IS PROHIBITED

One reason a retaliation prohibition is so significant in the legal realm is its relationship to part (b) of the *Faragher/Ellerth* affirmative defense.[16] If an employer does not include a no-retaliation provision in the policy or does not enforce such a provision, or fails to train on no retaliation, it may become problematic for an employer to establish part (b) of the affirmative defense. If an employee had a legitimate reason to believe that he or she would be retaliated against, it could certainly be seen as reasonable if the employee failed to take advantage of any preventive or corrective opportunities.

Clearly, a fear of reporting harassment would make the policy worth merely the laser paper it was printed on. Trainers should take the participants through a logical progression as to why this is the case, i.e., if an employee fears retaliation, the employee will not report it; if the employee does not report it, the employer will not know about it; if the employer does not know about it, harassment will continue. That is why trainers should emphasize that retaliation is treated as seriously as an incident of harassment and that employees should report incidents of retaliation on the same basis as an incident of harassment.

 TIPS:
- An easy way to define retaliation up-front is to describe it as simply "getting back" at someone. Once the importance of the retaliation provision is discussed, trainers should move on to discuss the types of retaliation that can occur and when an employee is protected from retaliation. For instance, not only are employees who make complaints of harassment protected, but anyone who participates in any way in a harassment investigation is protected from retaliation. In addition, even if an employee makes an "unofficial" complaint of harassment, the employee is protected.
- Also, discuss what retaliation can look like. Retaliation does not just include terminating or suspending an employee, but any adverse action taken against the employee—from reduction in monetary benefits to changed assignments to demotions. Ask participants to help you develop a list on an easel pad.
- Make clear that retaliation does not just mean managers and supervisors retaliating against a subordinate for complaining of harassment, but often involves co-employees, even co-employees who were not directly involved in the harassment complaint. Co-employees tend to take sides in a harassment "dispute," and this can lead to harassing behavior including threats, uncooperative behavior, physical actions, or the "silent treatment."
- You should make clear that the no-retaliation policy is the cornerstone of the harassment policy since without a policy against retaliation, people with legitimate complaints would not step forward because of a fear of reprisal. Emphasize that retaliating against someone is

treated just as seriously as harassment and therefore will result in discipline up to and including discharge, regardless of whether it is a first offense. Thus, in training, retaliation must be defined and the employer's commitment to no retaliation must be emphasized.

PROCESS FOR REPORTING RETALIATION

"Who do I call to report retaliation?" Explain that employees must report incidents of retaliation on the same basis as they report incidents of harassment and reiterate how an employee should use the complaint procedure to report any instances of retaliation. (However, if your policy has a separate reporting procedure for reporting retaliation, then follow that procedure.)

Special Obligations of Managers and Supervisors

Generally, the preceding content should be implemented for all audiences. However, if the audience includes managers/supervisors or high-ranking employees, a section on their special responsibilities under the policy should be included.[17] We use the terms "manager" and "supervisor" interchangeably to indicate a person with the ability to hire, fire, or direct employees' work. We recommend that you take a broad (and realistic) view of who is a manager or a supervisor in your organization and include not only those who have that title, but also those who might be perceived as managers or supervisors by employees making complaints. Clearly, the chief difference in content between training managers and supervisors and rank-and-file employees is the added responsibility management employees have with respect to the anti-harassment policy.

When determining the liability of an employer, the action or inaction of a manager or supervisor is paramount. If a manager or supervisor is aware that harassment is occurring but closes his or her eyes to the problem, there may be liability ramifications. Therefore, managers and supervisors must take prompt and effective remedial action at all times. For when a manager or supervisor is on notice of harassment, so too is the employer, and the clock begins to run on a reasonable response. Thus, it must be made crystal-clear that when a manager or supervisor hears, either directly or indirectly, about harassing behavior or observes harassing behavior, the manager or supervisor is on notice of harassment and must take prompt, effective action to stop the harassing conduct and prevent it from ever occurring again. This means following the harassment policy by addressing the behavior directly and/or reporting it to the appropriate official.

Training programs must focus on giving managers and supervisors the tools they need to successfully implement a policy. For the key to a successful policy is directly linked to how well management responds to and abides by the policy. Under the *Faragher* and *Ellerth* decisions it is now more important than ever for employers to train their managers and supervisors. Remember, inappropriate employee conduct is judged by the notice that the employer has received, i.e., once you *know or should have known,* you must act.[18] But for manager and supervisor harassment an employer can be liable for harassment whether or not it knew or should have known of the situation. Accordingly, careful selection, training, and monitoring of managers and supervisors in an effort to prevent harassment is often an employer's only defense.

Managers and supervisors, including human resources professionals and upper-level management, must shoulder much of the responsibility for keeping the workplace harassment-free, not only for legal reasons, but because their behavior will influence the actions of all employees. If managers and supervisors act in a harassing manner or allow employees to engage in harassing behavior, it could appear as if the employer tolerates harassment or other unethical or illegal behavior. In addition to the issues already covered above, when dealing with managers and supervisors the following points must be addressed:

1. Managers and supervisors must understand the full contours of quid pro quo harassment and their role in avoiding it. There is no affirmative defense for management conduct if quid pro quo harassment is proven. Therefore, managers and supervisors need to understand not only what constitutes explicit quid pro quo situations, but also implicit situations that can lead to claims of quid pro quo, including the dangers of dating and romantic involvements with subordinates. Although such consensual relationships are legal, they can easily lead to unlawful discrimination because co-workers of the subordinate may be able to claim that favoritism toward the subordinate denied them rights or benefits they should have received. Of course, a spurned lover can create a multitude of problems, both legal and otherwise. If you have a non-fraternization policy for management, this is a good time to remind managers and supervisors of that policy.

2. Managers and supervisors must be trained to immediately recognize harassment and sexual harassment or actions that will potentially lead to prohibited conduct, even when no one

has complained about the conduct. They must have a heightened duty to report potential harassment or discrimination. Therefore, managers and supervisors should be trained to immediately respond to any complaints brought to their attention by the employees under their supervision. In other words, they should know how to handle employees in a sensitive way when receiving complaints, whom to pass those complaints on to, and how to handle the situation in the immediate interim.

3. In our experience, most organizations do not want their managers and supervisors to conduct investigations of harassment complaints. This is usually a function that resides in the human resources department. Managers and supervisors must be made aware of that so they know whom to go to when a complaint comes to them. They also must know not to "investigate" complaints on their own without the assistance of Human Resources. (Of course, anyone who conducts investigations should be properly trained to do so.)

4. You may also want to warn managers and supervisors that if they are in a meeting with an alleged harasser, even in a nonunion environment, the alleged harasser may have the right to have a co-worker present—if he or she requests it. This results from the 2001 decision of the D.C. Circuit in Epilepsy Foundation of Northeast Ohio v. NLRB,[19] which extended "*Weingarten* rights" to the nonunion setting. *Weingarten* rights afford unionized employees the right to have a union representative present at an investigatory interview which the employee reasonably believed might result in disciplinary action. According to the court, upon request, all employees, whether union or nonunion, have the right to have a co-worker present during disciplinary meetings. We recommend contacting your favorite labor lawyer for more information and guidance on this new development.

Dealing with an actual complaint of harassment is the key point that must be addressed. Certainly, there are an infinite number of scenarios a manager or supervisor may confront when an employee seeks to report harassment. As an aside, because of time constraints, harassment training should not turn into a management skills class. There is just not enough time in harassment training for it to be also used to teach upper-level employees how to respond to problem employees or situations.

That said, however, managers and supervisors should generally approach each harassment complaint the same way. They should focus on the moment that they will have the most influence on how the harassment complaint will finally play out—the moment when the employee approaches the manager or supervisor about a harassment problem. Will it be taken care of professionally, promptly, and effectively or will it metastasize into potential liability and/or a lot of time and heartache for the employer? It is what we like to call the "critical moment mantra." Perhaps the simplest way to train managers and supervisors on handling this moment is to break the moment into two steps: the response to the employee and the resulting action. With respect to the response to the employee, the manager or supervisor must take detailed notes on the where, who, why, and when questions of the incident. But just as important, the manager or supervisor must provide an environment of openness and empathetic responsiveness. For example:

"I'm glad you came to me."
"We take complaints seriously."
"We will get back to you."

The resulting action simply means that a manager or supervisor must always contact Human Resources or follow whatever procedure the organization has determined. The manager or supervisor must also follow up with the complaining employee, regardless of the result of the investigation. Accordingly, managers and supervisors must be trained regarding their confidentiality obligations to a much greater extent than the general employee population.

TIPS: ■ "As a manager, when do I have to report what I see?" We often tell managers and supervisors that they need to do something about harassing conduct not only when someone has reported it, but when they "see it, hear it, or hear about it." Managers and supervisors unfortunately often have the mistaken idea that if no one comes forward to complain they don't need to do anything. Using examples here helps to bring the abstract down to the practical level. "You walk by the break room and hear Chris telling a racist joke or making a comment about a co-worker's disability. Do you know about inappropriate conduct? You bet!" Try some short scenarios where the participant has to decide if he or she is on notice.

■ You need to be clear about how and when your organization wants managers and supervisors to respond to harassment that they see. For example, do you want managers and supervisors to intercede when they see a situation that might be harassment, or do you want them to report it to Human Resources, or both? Our clients often require managers and supervisors to report all complaints made to them to Human Resources, but often permit a manager or supervisor to use his or her own judgment when witnessing an incident that might be harassment. This is particularly true if the situation is not severe and can be resolved with a statement like, "I'm sure you didn't mean anything by that remark, but it might violate our harassment policy. Please don't make those kinds of statements again." One suggestion: managers and supervisors should always document this kind of intervention, and further incidents should be reported to Human Resources. Again, there are several proper ways to handle these situations, but the key is to be fair and consistent and never ignore them.

∎ Managers and supervisors must be trained on how to respond to employees who try to report possible violations of the policy on the condition that the manager and supervisor promise "to keep it confidential." They should be trained to respond to such situations by saying something like, "I will keep this complaint as confidential as is possible. I can assure you that only those persons with a need to know will be told as we conduct our investigation."

∎ This is a great opportunity for a role play, particularly for a manager/supervisor session. Approach your co-trainer (or a participant that has agreed to participate) and begin to make a complaint. Start with: "As my manager, I wanted to tell you about something that I am uncomfortable about, but I want you to agree to keep it completely off-the-record, 100 percent confidential." Employees frequently want the harassment to stop, but are not willing to formalize it into a complaint or pursue it to a final decision. The employee usually doesn't want anyone to get in trouble or doesn't feel he/she should stick his/her neck on the line. The point that must be emphasized to both rank and file and managers and supervisors is that an investigation must always be made after a complaint of harassment. There is no such thing as an off-the-record conversation. Ask the manager group: "How would you respond?" Ask the employee group: "Any problems with this?"

∎ Managers and supervisors must also be trained to abide by the "need to know" rule themselves. In other words, managers and supervisors must be taught the importance of limiting their own dissemination of the allegations to only those persons with a "need to know." They should be taught who has a "need to know" and should be reminded that if they publicize the allegations outside the "need to know" circle, they can be subject to disciplinary action, up to and including termination of their employment. This is especially important because statements by managers and supervisors outside the "need to know" circle could trigger employer liability in a defamation suit.

Special Considerations for a Unionized Environment

While the content of the training remains the same, some of the processes and concerns of union participants are different. For example, union members may have "*Weingarten* rights," and under some collective bargaining agreements a union may co-investigate claims. It is important to incorporate the relevant union issues in your training to show that the policy does not contradict their rights, but rather is in addition to them.

TIPS: ■ Be clear about your policy's reporting procedure in a union environment. If it applies, remind the participants that if they are a part of the collective bargaining unit they can go to their union representative to report a complaint, but they should/may/must (whatever language is drafted) also report to the employer representative under the policy to let the employer know they are making a complaint so that the employer can take prompt effective action to end the conduct.

Special Consideration for the Public Sector Environment

In a public sector environment, there are a number of different considerations that do not necessarily change the essence of the training, but which affect surrounding issues. Also, since there are additional claims that can be brought,[20] there is an even greater need to do training,[21] as training can help a public entity show that harassment is not condoned and, in fact, actually violates its adopted policy and practice.

First, public employees have First Amendment and other constitutional rights that need to be taken into consideration.[22] Second, to the extent that you are training public managers about discipline for harassment violations, there should be built-in training reminders that public employees have due process rights and hearing rights that attach to certain positions. For example, tenured employees and those covered by collective bargaining agreements are generally found to have a "property" right in their positions and are entitled to a pre-termination hearing prior to discharge.[23] In addition, if a public employee is accused of harassment, that employee may be entitled to a "name-clearing" ("liberty interest") hearing before discipline is imposed.[24] Finally, public employees are frequently covered by unique collective bargaining contractual provisions that may affect not only discipline but also training considerations.

LESSONS LEARNED

To: One Canoe, No Paddle
From: The Trainers
Re: Lessons Learned

Dear One Canoe, No Paddle:

Here are some lessons learned on what should, at a minimum, be the content of a harassment training. It is your long-lost paddle! It will steer you toward calm waters and away from rough legal liability and the rocky shores of wasted time, low productivity, and heartache:

1. Distribute and then focus on your policy—not on Title VII.
2. Make sure you include not only sexual harassment, but harassment based on all protected characteristics, as they are listed in your policy.
3. Focus on conduct that is prohibited by your policy, both quid pro quo and hostile work environment, and be as specific as you can with examples of what would and would not violate your policy. But make sure participants know the list is not all-inclusive and that they must use their own judgment from the perception of others. Help them develop that judgment.
4. Give a lot of emphasis to your policy's own reporting process, including the variety of avenues for making complaints.
5. Be clear that the policy absolutely prohibits retaliation against anyone for bringing a complaint or cooperating in an investigation.
6. With managers and supervisors, focus on their affirmative obligation to report harassment.
7. Remember, you may have other policies relevant to the harassment training—for example, e-mail, Internet, code of conduct, and other such policies that you may want to discuss in the training. Go ahead. Do it. The more you enforce all the related policies, the stronger the message is.
8. If you are in a union environment, make sure your training addresses the special considerations related to it.
9. Reiterate at the end of the training that the employer has taken many steps to create a harassment-free workplace, including: establishing clear policies and procedures to prevent harassment, educating employees about the problem, specifying the procedure to use if a violation occurs, and actively encouraging open communication and complaints. But emphasize again that the employees are the key in the process of creating a harassment-free workplace. That it is the employees' responsibility to:
   • Know the policy;
   • Role-model the policy; and
   • Govern their own behavior.
10. You can do it!

Signed,

The Trainers

Notes

1. *See, e.g.,* Cadena v. Pacesetter Corp., 224 F.3d 1203 (10th Cir. 2000).

2. Look to see if you have a nonfraternization policy; it may govern workplace dating.

3. From a legal perspective, the "threat" or "reward" must be carried out to constitute a quid pro quo claim. But this information is not critical to the training of either employees or managers and supervisors. You want employees to report *anything* that remotely resembles quid pro quo, whether it is carried out or not. And managers and supervisors need to know to stay away from even the perception of a quid pro quo situation. By the time quid pro quo has become a threat or a reward, it's considerably too late.

For those of you employment lawyers who can argue with the best of them about the differences between quid pro quo, the threat carried out, and the threat not carried out, we want you to know that we understand the legal differences that attach to each. But, for training purposes, we don't get distracted by those differences because the conduct in either case is prohibited under the organization's policy.

4. For example, one online company in Manhattan is faced with two ten-million-dollar lawsuits filed by two women. One of the women, a software engineer, dated a senior VP and the other woman, an account executive, dated a second senior VP. Statistics may also hammer home the point. A study conducted by The Employment Law Alliance found that 70% of adults agreed that romance at work can cause favoritism and poor morale. A recent MSNBC survey of over 30,000 people found that 53% of women and 20% of men in the workplace had a relationship with someone higher up in the pecking order. *See also,* Scelta v. Delicatessen Support Services, Inc., 89 F. Supp. 2d 1311 (M.D. Fla. 2000) (a consensual relationship with a subordinate does not give a supervisor right to harass that subordinate); Nadeau v. Imtec, Inc., 670 A.2d 841 (1995) (employer had just cause to discharge a male employee who violated a written order to limit his contact with a female employee with whom he had previously had an affair); Lipphardt v. Durango Steak House, 267 F.3d 1183 (11th Cir. 2001) (affirming a jury verdict for an employee who was a former lover of the alleged harasser, her supervisor).

5. *See* Breeding v. Arthur J. Gallagher and Co., 164 F.3d 1151, 1158 (8th Cir. 1999); Wallin v. Minnesota Department of Corrections, 153 F.3d 681, 687 (8th Cir. 1998); EEOC Enforcement Guidance: Vicarious Employer Liability for Unlawful Harassment by Supervisors (6/18/99), EEOC Compliance Manual (BNA).

6. *See* Appendix G.

7. However, before you demonstrate the offensiveness of any example, you might want to remind participants of the ground rule that this may be offensive and that it is for training purposes only. *See* Chapter 6, note 3.

8. *See* Antle v. Blue Cross & Blue Shield of Kansas Inc., 75 F. Supp. 2d 1248 (D. Kan. 1999) (birthday card to plaintiff joking that "lots of people your age forget where they live now and then" leads to lawsuit); Anderson v. Aurora Township, 1997 U.S. Dist. LEXIS 20805 (N.D. Ill.) (birthday card stating, "Happy Birthday! I think it's awful that some people make nasty jokes about another person's age—fossil face" was evidence of age discrimination).

9. *See* statistics compiled by the EEOC at www.eeoc.gov/stats/harass.html.

10. *Faragher,* 524 U.S. at 787. This means if a woman is the victim, it will be considered from her perspective. *See also* Oncale v. Sundowner Offshore Services Inc., 523 U.S. 75 (1998) (objective severity of harassment should be judged from the

perspective of a reasonable person in the plaintiff's situation considering all of the circumstances); Ellison v. Brady, 924 F.2d 872 (9th Cir. 1991).

11. *Meritor,* 477 U.S. at 67.

12. *See* Dowd v. United Steelworkers, 253 F.3d 1093 (8th Cir. 2001) (offensive conduct does not necessarily have to transpire at the workplace in order for a juror to reasonably conclude that it created a hostile working environment); Bowers v. Radiological Society of North America Inc., 101 F. Supp. 2d 691 (N.D. Ill. 2000) (supervisor propositioning employee in supervisor's home was relevant evidence even though it occurred outside the workplace); P.F. v. Delta Airlines Inc., 2000 WL 1034623 (E.D. N.Y. 2000) (actionable claim against an employer whose employee was forced to work with a co-worker who had previously subjected her to an off-duty nonwork-related assault); Wildman v. Burke Marketing Corp., 120 F. Supp. 2d 1182 (S.D. Iowa 2000) (marketing manager can proceed to trial after suffering three days of harassing conduct during a Las Vegas convention trip).

13. *See, e.g.,* Swinton v. Potomac Corp., 270 F.3d 794 (9th Cir. 2001).

14. *See Meritor,* 477 U.S. at 73 (grievance procedure required employee to complain first to his/her supervisor; supervisor was the harasser in the case at bar); Wilson v. Tulsa Junior College, 164 F.3d 534, 541 (10th Cir. 1998) (process was deficient where it permitted employees to bypass a harassing supervisor by complaining to director of personnel services, but the director was inaccessible).

15. *See* EEOC Enforcement Guidance for examples of the detailed fact-finding questions that should be asked during an investigation.

16. Part (b) of the affirmative defense states that the plaintiff employee unreasonably failed to take advantage of any preventive or corrective opportunities provided by the employer or to avoid harm otherwise."

17. *See* Anderson v. Deluxe, 131 F. Supp. 2d 637, 653 (M.D. Pa. 2001) (employer must do more than rely on its self-made labels of hierarchical structure; the key question is whether the plaintiff reasonably believed that an individual was in a position to either stop the harassment or to inform the employer of its existence).

18. *See, e.g.,* Kornely v. Carson's Ribs, 2000 U.S. Dist. LEXIS 17722 (N.D. Ill.) (even if the employer's policy designated a channel for complaints, the plaintiff may withstand a summary judgment motion by presenting evidence that the employer was given enough information to make a reasonable employer think there was some probability that she was being harassed).

19. Epilepsy Foundation of Northeast Ohio v. NLRB, 286 F.3d 1095 (D.C. Cir. 2001).

20. For example, under 42 U.S.C. § 1983 (1976, Supp. Ill. 79); *See, e.g.,* Board of Regents v. Roth, 408 U.S. 564 (1972); Perry v. Sindermann, 408 U.S. 593 (1972); *See, generally,* SWORD AND SHIELD REVISITED, A PRACTICAL APPROACH TO SECTION 1983 (M. M. Ross, ed., ABA Press 1998).

21. In addition to strictly legal concerns, public employers must also be concerned about political realities and the public policy ramifications of all of their actions, including both training and especially the failure to train.

22. *Compare* Johnson v. County of Los Angeles Fire Department, 865 F. Supp. 1430 (C.D. Cal. 1994) (First Amendment protects firefighters' rights to have sexually explicit magazine in the workplace) *with* Robinson v. Jacksonville Shipyards, Inc., 760 F. Supp. 1486 (M.D. Fla. 1991) (presence of nude pictures in workplace violates Title VII).

23. Cleveland Board of Education v. Loudermill, 470 U.S. 532 (1985).

24. *See* Paul v. Davis, 424 U.S. 693 (1976); Bishop v. Wood, 426 U.S. 341 (1976).

4

Into the Blender—Hit Puree: Putting It All Together

Dear Trainers:

OK, I think I've got the idea of the content here. Really, armed with Chapters 2 and 3, it seems like I should be able to make the training accurate and complete, which I know is an issue for a *Faragher/Ellerth* affirmative defense. But now that I have only a few more days until my session, I need to know how I can put this together into something that will hold the audience's attention and help them to know our policy. Any ideas? Should I put in all the information you've given me so far . . .

Signed,

Into the Blender—Hit Puree

. . . .

Dear Hit Puree:

Whoa! Don't do that, what a mess! Let's look at this in a logical manner. You are facing what is for us the hardest part of the whole process—putting the content together in some way that engages the attention of your audience and brings them to a better understanding of the topic. That is the way we think about design. It is the process of taking the content (all of those topics that must be present for your training to be super-effective), distilling it down to a variety of teaching points, and hooking those teaching points up to interactive tools for delivery. It is not quite as easy as hitting the Puree button on your blender—but once you get the hang of it, it will seem that simple. Read on!

Signed,

The Trainers

Overview

We will be the first in line to say that we are not now, nor have we ever been, instructional designers—that is, experts in designing training. We are simply attorneys who have developed our own principles of training, not from a textbook, but from getting up in front of participants—succeeding and, oh yes, failing. We have experimented with our hits and misses, sought feedback at every turn, written down what proved effective and discarded what proved disastrous. It is from our experience of conducting harassment training across the country that we impart Three Simple Principles and Five Steps to Design. Here they are:

Three Simple Principles
1. Training Is Not a Presentation
2. Training Must Be Interactive
3. Do It with a Sense of Humor

Five Steps to Design
1. Know the Content
2. Focus on Your Audience
3. Choose Interactive Tools
4. Calculate Your Time
5. Line up your content to your teaching tools, consider your audience, the time available, and the level of risk you want. Hit Puree.

That said, let's proceed.

What You Need to Know . . .

Three Simple Principles

Principle Number One: Training Is Not a Presentation

Let's start with what training actually is, because we have seen a lot of programs that don't really pass the test of being *training*. On numerous occasions, we have been retained to do training for a client whose last "training" was a two-hour PowerPoint lecture in a darkened room. At the end of the two hours, when the lights came back up, the "trainer" asked if the audience had any questions. They didn't. One year later, swamped with harassment claims, the client feels that his employees didn't "get the message." What a shock.

Well, what exactly *is* training? One dictionary defines training as, "making proficient through special instruction and drill." What does this definition tell us? Simply that training is not just instruction and should not be merely a lecture or a pedantic exercise. There must be another, crucial component—"drill." Or to use a different term, there must be a practice component to the instruction. Evolving one's thinking from lecturing an audience to training with participants should be the key focus when designing a course. It is this "drill" or "practice" that must be the grounding focal point of training. For it is the "practice" that breathes life and memory into the world of harassment training, making the audience feel like participants and not just a group attending a lecture.

That is our first principle: to have training, you must have instruction and practice. If you are standing up and talking to the audience for two hours, you are giving a presentation. That's fine. Maybe. But it is not training. In training, you must move from the abstract to the specific and let participants "practice" what you have told them. In interactive training, you have participants as opposed to an audience because the training class participates. If you are turning off the lights and showing a video for an hour or lecturing with beautiful and well-thought-out PowerPoint slides, well, that is simply not training. Practice, practice, and more practice makes training.

Principle Number Two: Training Must Be Interactive

Here is why training needs to be interactive: in order to get effective practice, you need interaction. Unless you have found a way to drill a hole in people's heads and pour in the content you want them to know, you are going to have to let participants "practice" what you are "preaching." This interaction requires "tools"—tools like role plays, hypothetical scenarios, exercises that allow participants to work together and alone to come to the answer on their own. In Chapter 5, we will discuss the types of tools that best generate interaction when training adults.

Learning for adults generally occurs when they feel a need to know something. When this need is communicated and accepted, adults will take responsibility for their own learning. In our experience, participants have a higher degree of acceptance of your message when they come to it on their own. We learned this from a great master trainer, Bob Pike, who says, "People buy their own data."[1] Unlike children, who may accept a concept because the teacher says it is so, adult learners can be skeptical of you and your message. Developing accurate and complete content with a toolbox of interactive exercises will go a long way toward ensuring success.

Also, remember that we adults tend to have very short attention spans, perhaps as a result of our acculturation to television and the thirty-minute sitcom. The use of interactive tools helps to create a climate conducive for learning and practice, a climate that motivates and engages adults for the entire training session. In addition, these interactive devices, along with the actual training environment, should appeal to all the senses, from bright colors in the exercises to music during downtimes, to physical activity during team activities. It is imperative that the session not be heavily weighted in any one area, as the monotony of engaging only one sense often will lead to boredom. We try never to do one segment for more than 20 minutes. Otherwise, no matter how fascinating the training is, eyes begin to droop, attention begins to wander, and posture begins to slump.

Principle Number Three: Do It with a Sense of Humor

One final guiding principle should be remembered when designing harassment training. That is the need to have fun and to have a sense of humor. That is not to say that you can make fun of the topic, the participants, or the policy. Of course not. Clearly, the issue of sexual harassment/discrimination is a serious and problematic one in today's society, and this must always be recognized. No one should get the impression that these issues are being taken lightly by anyone involved in the class. (See Chapter 7, "What Could Go Wrong?" for a further discussion of using humor.) On the contrary, the training must convey a sense of urgency and importance about the pervasive problems of harassment and discrimination. But that does not stand in opposition to having fun and having a sense of humor. Indeed, when one deals with difficult issues that are emotionally loaded, it is often helpful to have an appropriate sense of fun and humor about such issues. With respect to harassment training, this implicitly conveys the message that the training goal is not to point a finger at anyone; that the training is not being conducted because the employer employs bad people or believes each participant is a likely candidate for being a harasser. Instead, having a sense of fun and humor conveys the message that the participants are respected and will be treated like adults in this training session because adults have the ability to separate the humor from the seriousness of the issues.

Promoting an atmosphere of fun also helps to create a comfortable learning environment that makes it easier for trainers to receive the full participation needed for the interactive exercises to work properly. An environment of gloom and doom does not lend itself to a fully functional interactive experience, but instead promotes a lecture/pedantic environment that is the antithesis of adult training. And there is a very selfish reason—it's just much, much easier to do training in a fun environment!

Put another way, the goal of having a sense of fun and humor is to acknowledge our joint humanity while not disparaging the importance of the issues. How does one stay on the right side of that fine line? There are no easy answers. Finding that line comes with experience in the social setting and an inherent ability to carefully monitor one's audience and the audience's predilections—of knowing at what moments lightheartedness will prove effective and when seriousness is the order of the day. Because our recommended interactive exercises lend themselves to a sense of fun and humor, we find this atmosphere will arise naturally.

With our lawyer hats on, we need to say one more word about the three simple principles listed earlier—(1) practice, (2) interaction, and (3) humor. Not only do they make sense from a training perspective, but they also make sense from a legal perspective. More and more frequently, managers and supervisors are being deposed and asked about the training in harassment issues that they have received. And often, they do not remember anything about it. We often think of the story told to us by one of our partners, who described a manager being deposed in a harassment claim. The plaintiff's lawyer asked if he had been to training. He responded affirmatively. But when asked what he remembered about the training, he paused for a long moment and finally replied, "Lunch." Oh well. At the very least, using your training successfully for an affirmative defense requires that you strive to make the training interesting and, hopefully, more memorable than the lunch. Using learning devices, props, and interactive tools will help build that memorability, which we will discuss in Chapter 5. With these guiding principles in mind, let's move forward to the five steps to design.

Five Steps to Design

Step 1: Know the Content. This is the meat of your program. The rest is dessert.

So go back over Chapter 3 and review the content.

Step 2: Focus on Your Audience. Know what your audience thinks is important.

Who is the audience? That is the first question that we ask ourselves when we begin to put the content together with a design. Generally in harassment training, we are looking at one of three or four different audiences: (1) employees, (2) managers/supervisors, (3) sales representatives, and (4) executives. While some trainers may wish to combine the audiences for logistical or other reasons, we have found that these class groupings work effectively.[2] While the content of the training will be similar for all four groups, the emphasis and direction of the training may be somewhat different.

Apart from the global goal of eliminating harassment in the workplace, each audience has a different reason for being trained, and a different need to know the information. Start your design process there. When conducting training, a key to success is that the audience understand the importance/significance of the training. If that is not readily apparent, the motivation and attention of the audience will most likely prove lackluster at best.

So ask yourself: What does the audience need to know? Why is it important to them? For employees, who are generally not liable for harassment being committed at the workplace, the reason to learn about anti-harassment does not necessarily have a legal liability component.[3] But there are consequences. As such, the message to employees may focus on the employment consequences of engaging in impermissible harassment, and the comfort and lack of worry when you know exactly what the rules of the workplace are.

What do managers and supervisors need to know? This may center not only on their job responsibilities, especially as it relates to reporting procedures, but on the potential for individual liability. Many states permit manager and supervisor liability in their fair employment practice laws. When dealing with executives, the training may additionally emphasize business concerns—the risk of liability and the potential liability costs.

TIPS:
- We try, whenever possible, to focus on the positive benefits of understanding the issues of harassment in the workplace. Participants are usually more receptive to this approach than to "scare" techniques. However, we are not above letting participants know that there are serious consequences to being involved in harassment. Learning that the consequences can be as catastrophic as discharge does get some folks' attention.
- There are other questions about the audience that you also need to know, which concerns the nature of the workforce and the type of work performed. Is it a blue- or white-collar environment? Is it a service industry or manufacturing? What is the audience's business? All these questions must be answered before designing a course. Giving a blue-collar manufacturing group examples or hypothetical situations that occur in a white-collar service industry will not have the same desired effect as providing examples or hypothetical situations that specifically relate to the audience's day-to-day activities and culture.

■ For those of you who will be training at different organizations, one tip that seems obvious, but is often overlooked, is to learn as much as you can about the organization, the organization's competitors, and the cultures within the organization. Tailor your training to issues that are relevant to the organization. The more specific your examples and analogies are to the organization, the better the reception. Not only will participants enjoy the use of such examples and illustrations, but it will enhance your credibility, not as a "hired gun," but as someone who is part of the extended family and familiar with the participants' day-to-day concerns. For example, we have done quite a lot of training within the automobile manufacturing industry. We have learned that what works in stamping plants does not necessarily work in assembly plants. Same content, two different cultures. You can't change the required content, but you can adjust the focus and the approach of your interaction in role plays, case studies, etc.

Step 3: Choose Interactive Tools. Range from low to high risk for good interaction.

Let's say you are now convinced that the most important element of a training program is practice, and the most effective way to make training practical is through interactive training and tools. You will proclaim that interaction makes the training more engaging and lively for the audience. And you know that since interactive training is a two-way street, it is imperative that the audience achieve a comfort level that will permit them to fully participate in the process. The million-dollar question is: **How in the world can I make them participate?** We think this is why so few trainers try to be interactive. Call it a fear of rejection (you invite participation and nobody comes!). Or fear of silence (looking out over all those blank faces with eyes averted). The end result is the same—nobody talks, trainer gives up.

Here is a key to erasing your fears: the way to achieve interaction is to slowly build audience participation. Start with low-risk tools—that is, exercises that provide little difficulty for the participants in terms of both the content of the answer and the degree of psychological risk needed to answer. Here is what we mean by that. Suppose in the beginning of the training, you pose a question: "What is the most important thing that we need to understand about harassment in the workplace, and by the way, if you know the answer, please stand up and tell us." How many responses will you get? In our experience, none. Why? Too high-risk. There is no definite answer to the question and even if we thought we knew the answer, would we stand up and address the class in the first five minutes of the training? Not us!

What if, instead, we asked, "How many of you have ever read anything in the newspaper about harassment? If so, raise your hand." (And, here is an important touch: you raise *your* hand too, to demonstrate the results.) How many responses will you get now? That's right, a lot. Almost everyone has read something about somebody being sued (a "no-brainer" and therefore low-risk), and raising our hand is also a low-risk way to participate. As the training progresses, risk can be increased and you will find participants more than willing to speak out. (And then you can move on to fearing an *excess* of questions; see Chapter 7, "What Can Go Wrong.")

Another way to achieve a comfort level for the participants at the beginning of the training is to have participants work in teams. By forming teams, participants will not have to be individually responsible for exercises, but instead will have "safety in numbers." Teams promote social interaction and again make it easier for trainers to conduct an interactive training session. Think of it this way: you might not be willing to speak in front of the entire group in the first 20 minutes of the training, but you might feel okay about introducing yourself to three or four other team members and discussing a question that has been posed. When you allow participants to get their "voice" early, there is less resistance to the messages and much more engagement in the learning process.

Therefore, as you build your training program, you need to think about the risk level of the activity and progress from low risk to high risk as you move through the program.

Remember too, you must be comfortable with the risk level of the activity you take on as well. Some trainers feel confident and comfortable with higher risk and others don't. It doesn't matter. Do what you are comfortable with and the participants will be comfortable (and engaged) too.

Step 4: Calculate Your Time.

You need to know how much time you have with each of your groups. Generally, the most common training we do is three hours for managers and supervisors, two hours for employees, and one to three hours for executives. There is no magic in these numbers, but we know that we can get all of the content for each of these groups into those time periods. You may have more, you may have less, as the operational demands of your business may govern the time you can spend.

Once you know the amount of time you have, with the start and finish times, you can build in breaks. Our rule of thumb is: at least one break every 90 minutes. Therefore, generally do one 10–15 minute break in a three-hour session. In a two-hour session, we may do a shorter break after an hour, or more commonly, no break at all. If we go for more than two hours, we always give a break.

Step 5: Line up your content to your teaching tools, consider your audience, the time available, and the level of risk you want. Hit Puree!

Before you hit Puree and put all the pieces together, however, you should know the type of interactive tools that are available. So let's move on to Chapter 5 and take a look at the vast array of teaching tools that can be utilized.

LESSONS LEARNED

To: Hit Puree
From: The Trainers
Re: Lessons Learned

Dear Hit Puree:

You are coming into the home stretch. Remember:

Three Simple Principles

1. Training is not a presentation;
2. Training must be interactive; and
3. Do it with a sense of humor.

Five Steps to Design

1. Know the content;
2. Focus on your audience;
3. Choose interactive tools;
4. Calculate your time; and
5. Line up your content to your teaching tools, consider your audience, the time available, and the level of risk you want.

Then all you need to do is hit Puree!

Signed,

The Trainers

Notes

1. ROBERT PIKE, CREATIVE TRAINING TECHNIQUES HANDBOOK (Lakewood 1994).

2. In our experience, executives sometimes go through the same program as managers, and sometimes are mixed in with the manager group. This is often a "culture" or logistics issue.

3. Be aware that in California there is individual liability for employee harassment. CAL. GOV. CODE § 12940(j)(3).

5

Juggling with Fire:
Interactive Delivery Tools

Dear Trainers:

I am worried that while my training will be legally accurate, it will be boring! What can I do to keep the participants from writing on their evaluations that they were BORED or—worse yet—found it less painful to go to the DENTIST!! Should I try juggling . . . with fire?

Signed,

Juggling with Fire

. . . .

Dear Juggling with Fire:

"No" to juggling fire! It is not interactive enough unless you juggle with the audience's involvement. However, you are on the right road in thinking about how you should deliver the teaching points in a fun and interesting way that involves the audience. There are many reasons for making the training inter-active and there are many successful methods trainers use to deliver concepts on the law and workplace policies. Read on!

Signed,

The Trainers

Overview

Imagine that your growing-up years were one giant long lecture by your parents about how to properly do "this and that" until you turned of age to live on your own. Yes, for almost two decades you sat there doing nothing but hearing about the skills you would need as an adult. Then, one day the lecture ended. You were sent into the real world where you needed to use the skills they endlessly lectured you about—reading, writing, driving, taking care of your own clothes. . . . Would you have mastered those skills? No, not without personal practice! And that is what effective training is all about—the opportunity to learn new concepts and personally practice them to acquire new skills. Luckily for trainers, they have many different methods and tools to pick from when deciding on how to make their training interactive.

If our previous chapter did not persuade you to ditch the lecture methodology, how about a few statistics to back up our point. According to one study, students were attentive to the information only 60 percent of the time in lecture-oriented classes.[1] This gives new meaning to "in one ear and out the other." And, if an employer is paying to send employees to training, this gives new meaning to "flushing money down the drain" when you consider 40 percent of the information goes with it. Of course, there are other drawbacks to presenting information solely by lecturing, including the decrease of audience attention as the clock ticks. Studies have also reported that lecturing appeals only to auditory listeners and misses connecting with visual learners, and that many participants simply don't enjoy learning that way.[2] In contrast, just adding visuals to a lecture causes a dramatic increase in retention, from 14 to 38 percent.[3]

Training is not a presentation nor is it a facilitation, it is somewhere in the middle. In a presentation the speaker imparts the content to a passive audience. In a facilitation, the facilitator has no content but rather facilitates the group's process. In training, however, the trainer has content *and* a process and turns the audience into participants who are charged with sharing in the learning process to acquire skills.

Confucius had it right in 451 B.C., and it has never been said better. "What I hear I forget; what I see, I remember; but what I do, I understand."

So by now you should be sold on putting your pointer stick in the garage sale. This chapter will take you deeper into the land of props, tools, and methods that trainers can successfully employ when conducting harassment prevention training. For your convenience, we have also rated the tools in terms of their risk level from the perception of the audience. As mentioned in Chapter 4, low-risk activities mean participants likely will not feel threatened by the interactive tool; medium-risk activities mean if participants are shy they may not volunteer at first, but if other outgoing individuals get involved, they may find it safe to follow; and high-risk activities mean the spotlight is on the participants and there is a possibility they could lose face. As you will see in harassment training, the most successful tools are low-risk and medium-risk activities, and that's what we have provided you with. A note on group size: not every activity will work with a large group. But don't succumb to the tendency to lecture just because your group is large and seated in an auditorium. Look through this chapter and see what you can adapt to your situation. Our favorite large group activity uses communication cards. Try it, you'll like it!

What You Need to Know . . .

Props

Communication Cards [Low-Risk Activity]

The image of props conjures up things on a stage like a telephone, a bowl of fruit, or a sofa. But in "trainingland," props are used to keep the audience involved. One way to foster fun communication between participants and the trainer is to give each member a set of "communication cards," which are used to answer questions posed by the trainer. We use "Yes/No" and "True/False" cards in almost every training. Each card is the size of a large index card or a half-sheet of paper and is color-coded green for positive responses and red for negative responses.

Why? Well, think about what happens if a trainer asks the group, "Is it okay to send sexual e-mails to co-workers while at work?" The group usually just sits there staring back at the trainer fearful of answering first and even more fearful of answering wrong. Even if the trainer is lucky enough to have one person raise their hand or another shout out an answer, it is typically only after some time has elapsed and some kind soul begins to feel sorry for the trainer. In any case, prolonged silence in the room is an opening for boredom to set in. However, with "Yes/No" communication cards the audience is able to flash their answer without taking the risk of looking foolish in front of the group. This works because the trainer also holds up the two cards, waves them around, and then instructs the participants to "talk to me" by using the cards. It is more fun, a lower-risk activity and keeps the training moving.

The trainer also can use communication cards that read: Legal/Illegal, Okay/Not Okay, Policy Violation/No Violation, Report/Don't Report, Good Practices/Bad Practices . . . you get the idea.

TIPS:
- Generally, use one or two sets of cards per course so that participants don't have numerous sets in front of them to pick from. If you use two sets, then each set should be used for a different exercise. For example, if you are reviewing a True/False pre-test, then have the participants use the True/False cards. If in another exercise you are asking participants to judge conduct or situations under your harassment prevention policy, then use the Okay/Not Okay cards.
- The trainer should keep the cards in hand and use them to demo the interaction as well as answer the questions!
- The card sets work best if there are one or two words per card and they are color coded in green and red. For example: Okay (green)/ Not Okay (red); True (green)/False (red); Legal (green)/Illegal (red), and so on.
- This is a great tool to use with medium to large groups. The cards create interaction that is manageable no matter how many faces are staring back at you from the audience!
- **Done wrong:** Tell the participants to use their "Report/Don't Report" cards and then ask them, "Does having a sexy calendar at your desk violate our harassment prevention policy?" Obviously, those communication cards cannot properly answer the question and everyone will be frustrated. The better question is, "If you see a sexy calendar on someone's desk, should you report it or not?" With this question, the participants will be able to select a card and "talk" to you. Remember, pose your questions so the communication cards are responsive.

Interactive Tent Name Cards [Low-Risk Activity]

Tent name cards are a multipurpose prop. They serve as a simple, inexpensive way for participants to display their names and participate in an interactive exercise. Unlike pre-made plastic name cards, the tent name cards are completed by participants at the beginning of the training to get them involved and are flexible enough to accommodate the unscheduled participant who otherwise would not have a pre-made name card. Because participants fill in the tent name card at the beginning of class, it helps engage them from the moment they walk in the door.

A tent name card is simply a colored piece of paper folded in half and placed in front of the participant like a tent. On one side there is a line with a prompt under it, such as "Print your name on the line above." This side, the front side, faces the trainer, making it easier for the trainer to call participants by name to create a more friendly interactive learning environment. The back side of the folded card faces the participant. This side can serve two purposes. On the right side you can list "ground rules," such as no legal advice, no discriminatory remarks, no debating the policy, and so on. (see Chapters 6 and 7). This helps the participants remember the ground rules throughout the training. On the left side you can have a nonthreatening interactive exercise that gets debriefed sometime during the training. For example, you may want participants to complete a list that helps you get to know them better without creating liability, such as "List three reasons why it is a good idea to have a harassment prevention policy at work." Or, you may want to ask a thought-provoking question like, "What kind of problems could employees experience if an employer did not have an anti-harassment policy?" This interactive exercise is useful not only because it engages the participant at the beginning of the training, but because people "buy their own data."[4] In other words, participants are more likely to be persuaded by their own responses than by ours.

 TIPS:

■ The trainer should "meet and greet" participants by sharing his/her tent name card and reading the participants' completed cards to initiate a friendly, interactive climate. We always have operated on the assumption that the training begins when the first participant walks in the door.

■ The trainer must complete his or her own tent name card before the training begins so that the participants can see what the final result looks like. This includes completing the interactive exercises. There is nothing more "disconnected" than a trainer asking a participant to complete a tent name card and then just holding up a blank one that the trainer never took the time to fill out.

■ Using multicolored paper for the tent name cards makes the room look cheerful and sends a message to the participants that the training will be fun.

■ The best type of paper to use is "card stock" because it is heavier than standard copy paper and stands up better on tables. Card stock is generally only pennies more than regular paper and well worth it.

■ The tent name card can also be used for many other purposes. For example, it can be designed to have a cut-out or perforated "wallet card" with the reporting procedure on the front so that at the end of the training session the participant can tear out the wallet card before the tent card is thrown away. Other uses include having a job aid (such as a graphic drawing of the reporting procedure) on the inside of the tent name card so that participants can use the tent name card after the training.

■ **Done wrong:** Never ask participants to reveal personal information in an interactive exercise or on a tent name card because it could feel threatening to the participants, chill participation, and possibly create liability. For example, don't ask, "Have you ever joked around with another employee because of their age, race, or disability?" Moreover, such questions could cause "notice of harassment" issues or legal liability (see Chapter 7), or be seen by participants as a trick question that could destroy the trainer's relationship with the audience.

Nouns, Music, and Metaphors [Low-Risk Activity]

We are strong believers in using common metaphors that all participants can relate to in the training. For example, we talk about the employer's harassment prevention policy being similar to the house rules participants had growing up or have in their own homes today. Everyone can relate to house rules. And even if participants say they did not have any house rules growing up, they really did. Everyone had to lock the front door, put away the dishes, or clean their room. Such metaphors help audiences relate to legal concepts without getting into the mundane matters of why we have corporate policies. Yawn!

There are many other fun props that can make the training interactive and memorable—for example, using a man's shoe and a woman's shoe to represent the concept that conduct is perceived from the gender-specific shoes of the listener, not from the intention of the speaker. Virtually any noun can be used as a prop. Dice can be used to signify risk, and back-scratchers can illustrate quid pro quo. Look around and see what's in your office, home, or party store that can liven up your training.

One of the greatest climate-setting props of all time is music. It creates moods, calms nerves, and blows life into deadly silent rooms filled with strangers.

TIPS:
- Play upbeat music while participants enter the room. This sends a message that this training is going to be fun, interactive, and lively. Not to mention different![5]
- Play soft background instrumental music (non-vocal) while participants are engaged in reading activities, like reviewing their anti-harassment policy or reading hypotheticals.
- Play fast rhythm music when participants take breaks and come back from breaks. This keeps the momentum of the training going while everyone is moving in and out of the room.
- **Done wrong:** Playing vocal music while participants are required to read the harassment prevention policy. Many people cannot concentrate while singing is going on in the background, even softly, even Aretha Franklin's "R-E-S-P-E-C-T"!

Tools

Handouts [Low-Risk Activity]

Harassment training involves unique circumstances when it comes to handouts or written materials because they may end up as attached exhibits in litigation and evidence at trial. Be careful not to let them come back to haunt you. Rather, draft all handouts to support your efforts. Trainers call this the "handout rule."

TIPS: ■ Since a judge or jury may be handling the handout, it is a good idea to use it as an advertisement. Go ahead, be creative. For example, add the employer's reporting procedure at the bottom of handouts or a reminder about the employer's zero tolerance for harassment. Use it as a billboard to send the subliminal message that you are using reasonable care while overtly imparting information to the participants. These types of messages work great on pre-tests, case studies, role plays, tent name cards, and anything that is given to participants to keep or work with.

■ **Done wrong:** Create handouts which are poorly phrased, have difficult choices, and provide a scoring scale of pass/fail. We have seen trainers use such handouts. Unfortunately, we have also seen attorneys get a lot of mileage at trial from such handouts, too!

Pre-Tests [Low-to-Medium-Risk Activity]

Contrary to how we felt about final exams growing up, pre-tests can be used effectively in training. They should be fun, informative, and not tricky. For example, a ten-question True/False questionnaire given to participants before the training helps them to focus on the issues they are about to be trained on. It should address all the key concepts in the training such as prohibited conduct, reporting, and retaliation. In addition, while it is designed to be completed before the training begins, it can be debriefed at the end of the training as a recap of the course. Reviewing a pre-test at the end shows participants how much they have learned.

Because of the handout rule, consider eliminating the use of a writing tool to record answers. Instead, use communication cards. Or, at the very least, have the participants review the pre-test for the correct answers at the beginning of the course, but wait to fill it out until the end of the class as part of the final debriefing exercise. This method is a win-win for the participants and the employer. It allows the participants to use the pre-test as a resource with correct answers after the training has ended, while at the same time not creating any "bad evidence" to haunt the organization at trial.

TIPS: ■ The pre-test can be used in addition to the tent name card so that participants are engaged in lots of interactive activity even before the training officially begins.

■ The pre-test should incorporate statements about your employer's anti-harassment policy so they can be reinforced again at the end of training.

■ Use a title for the pre-test that won't make participants apprehensive. For example, seeing "All About Our Harassment Prevention Policy" at the top of the page is more user-friendly than "Pre-Test." Or you could label it a "Questionnaire" and bypass the flashback for those who feared tests growing up.

■ When debriefing the pre-test at the conclusion of the training, use the interactive True/False communication cards again to keep the training interactive up to the last second and avoid creating evidence that can be used against you in court.

■ **Done wrong:** Designing a pre-test with a choice of four answers makes the test too long to complete and slows down the beginning of the session. It also makes it too complicated to debrief at the end of the training when participants are tired, want to finish up, and don't have the energy to flip through four cards A, B, C, D. True/False is the fastest and easiest way to go!

Power Points or Slides [Low-Risk Activity]

A PowerPoint or slide show is a lecture with pictures. While this is better than a straight lecture without visual aids, it is even more interactive if you can have the audience respond with communication cards. For example, if the True/False pre-test that the participants completed is also on a PowerPoint presentation, you will be appealing to auditory adult learners (by lecturing) and visual learners (by PowerPoint).

 TIPS: ■ Make the PowerPoint interesting, but do not count on your computer to always work. We wish we had a dollar for every time technology has failed us! That is the key drawback to relying on anything that needs to be plugged in. Therefore, have backup handouts and write the information on large easel pads—just in case.

■ **Done wrong:** Read the PowerPoint slides to your audience one after the other. You will be back in "lectureland." Drone on!

Video Vignettes [Low-Risk Activity]

Video vignettes are quite useful as a tool to help participants spot harassment issues. When buying, using, or creating video vignettes, there are a number of things to keep in mind:

■ To the extent possible, the vignettes should be in an environment that mirrors the participants' environment.

■ The vignettes should not be used strictly for entertainment purposes, but should incorporate teaching points for participants to debrief.

■ To touch upon as many teaching points as possible, a number of two- to three-minute vignettes may prove more successful than fewer, longer vignettes. Participants tend to forget the issues if the vignettes are too long.

■ Make sure that the character names are clearly articulated in the video so you can discuss their conduct in the debriefing. There is nothing worse than trying to discuss the conduct of a character who has to be identified every time as "the man in the blue suit."

■ Most important, to promote discussion, present situations that are not blatantly harassing. Instead, attempt to use situations in which there is some legitimate motivation for the harasser or where the harasser legitimately does not perceive his/her behavior as inappropriate. This not only mirrors reality, but helps participants to be able to see both sides, and develop judgment about risks and consequences.

■ Depending on the vignettes, disclaimers may need to be made to the participants before viewing, i.e., a warning that the language or situations in the video may be offensive and that the video is used only for training purposes and does not reflect the attitudes of the employer.

■ Try to find vignettes in which the harasser is not always male and the victim is not always female. It is helpful for participants to "see" this so they can recognize that harassment is not always directed at women by men. If you do not have a vignette showing this, explain in the session that while the vignettes show the majority of harassers as males and the victims portrayed as females, it is only because statistically the majority of sexual harassment complaints are made by females. However, any harassment violates our policy.

■ Never turn the lights completely off when showing a video vignette so that participants are not tempted to close their eyes and nod off.

 TIPS: ■ Before showing the vignette, prompt participants with a question or a "heads-up." For example, "You are going to be seeing a video for the next few minutes. Everyone on one side of the room, look for conduct that would be OK under our policy, and everyone on the other side look for conduct that is not OK." This focuses everyone's attention on the task and provides for a robust debriefing discussion, while utilizing communication cards to enhance the experience.

■ **Done wrong:** Watching the video vignettes without asking a question beforehand or failing to use communication cards during the video makes the experience too passive. When you flip the lights back on, you may even find a few folks dozing.

Hypothetical Situations [Low-to-Medium-Risk Activity]

Developing written hypothetical situations allows a trainer to tell a story filled with relevant teaching points. Thus, the story should involve issues that the group can debrief to reinforce concepts, spot issues, or practice skills. Hypotheticals should paint a scenario that is rich with information, but not so long that you need a road map to understand the story. Trainers should also design a series of follow-up questions that bring out the teaching points set forth in the hypothetical.

TIPS:

- Hypothetical situations should be printed for each participant and also can be read to the group. The follow-up questions can be given to teams to work on to present to the group. Remember, the best way to have the participants demonstrate to the trainer they have mastered the materials is to have them explain it to others.

- Another way to work with hypotheticals is to create one per team (or table group), representing different teaching points (conduct, reporting, retaliation) within the framework of one topic. Have each group work through their hypothetical, and then present back to the group. Remember, people buy their own data, so this is another way to have them own the information.

- **Done wrong:** Develop a hypothetical that is too long, takes forever to read, and is confusing.

Scripted Role Plays [Medium-to-High-Risk Activity]

Role plays are dramatic portrayals of people in various real-life situations. They are a great way for adults to apply new skills and can be entertaining for a group to observe. There are many different ways to structure role playing—from having the parts be scripted readings prepared ahead of time (a lower-risk activity) to improvisational, where the participants are given a general situation and fill in the details themselves as they interact (a higher-risk activity). However, remember that adults do not like to be embarrassed or feel incompetent in front of others, so you may want to consider using the role plays having the lowest risk—those done by the trainer and a willing, prepared participant or a co-trainer.

Since it is critical in harassment training that no stray or discriminatory remarks are made, scripted role plays are the safest way to keep control over the training while using dramatic tension to unveil teaching points. For example, a scripted role play involving an employee reporting harassing conduct to his or her manager/supervisor in which the manager/supervisor tries to talk the employee out of reporting is a useful tool to observe "bad practices." Another role play can demonstrate the manager or supervisor properly taking in the complaint as a live example of "best practices."

Role plays need to be debriefed after they are finished. Just as in debriefing video vignettes, the participants should be asked to answer specific questions as a table team. For example, "What did the manager/supervisor do right? What could the manager/supervisor have done differently? What consequences could the manager/supervisor suffer for trying to talk the employee out of making a complaint?"

TIPS:

- Make sure to have the same number of scripts as the people in the role play so that everyone gets their own to read from rather than distracting the audience by sharing scripts and turning pages.
- Have the debriefing questions up on an easel so that partners or table teams can debrief the scripts immediately after the role play has ended.
- **Done wrong:** The trainer creates a role-play situation where an employee is challenged with displaying new skills in front of her immediate manager/supervisor. The employee fears she is being evaluated by her manager/supervisor and, therefore, is reluctant to take on the role play or does so hesitantly and participates for only a brief time. The role play is a flop and other participants fear they will be called upon next to be made a fool of. We call these "fishbowl role plays" because everyone is watching and the hook is being lowered.

Scavenger Hunts [Low-to-Medium-Risk Activity]

Scavenger hunts are fun ways to interact with participants when they have to read an anti-harassment policy, e-mail policy, or code of conduct. The trainer prepares four to six questions relating to the policy. The individual or group then has to hunt down the correct answers that are located in the policy. For example, "List three kinds of conduct prohibited by our policy." "If you are an employee, whom should you report harassing conduct to?" "What is our policy about retaliation?"

TIPS:

- Set up a competition between one side of the room and the other to see which side finishes the scavenger hunt first. Most people enjoy a good competition, especially if there is nothing to lose, and it can move a low-energy exercise along.
- This is a good exercise for instrumental music to be playing in the background.
- **Done wrong:** Make the questions too complicated (one concept at a time, please), or where you cannot find the answers in the policy. Don't use too many questions. With this tool a little goes a long way.

Work Scripts [Low-to-Medium-Risk Activity]

Work scripts are a simple but helpful way to make sure employees have strategies and verbal skills to handle various situations. Generally, they are a three-step prepared statement or response that is easy to remember. While it is not necessary to remember the exact words, they should be simple enough for the user to remember when they need the "script."

Work scripts are a critical tool to give participants and can even drive a culture change in an organization. For example, an organization may want to train all its managers and supervisors to learn a simple work script to say when employees step up and make harassment complaints. "Thank you for coming. We take these situations seriously. Someone will get back to you right away." It is better to teach all managers and supervisors to be consistent in their responses. Such a unified response demonstrates the organization's commitment to a harassment-free environment. And, if litigation ever occurs, it looks better in a deposition transcript than other unprepared responses such as, "Don't complain to me, you nutbag, you are too sensitive!"

TIPS:
- Work scripts work well to teach employees how to apologize for unwelcome conduct without creating more tension. For example, "Thanks for letting me know." Or, "I am sorry if I offended you. It will never happen again. I apologize."
- Work scripts also work well in role plays involving a secret fact (see Simulations below) because participants have a chance to practice the scripts.
- Work scripts can be copied on the inside of tent name cards or on a wallet card as a take-away at the end of the training.
- **Done wrong:** Using hard-to-remember work scripts with difficult words or concepts. For example, "Thank you for coming to me because I am committed to an environment free from harassment. I know that you too are committed and that our mutual awareness will . . . yadda, yadda, yadda." This is too long and difficult to remember! Simply, keep it simple!

Simulations [Medium-Risk Activity]

One special type of role playing that deserves special mention and works well in harassment training is a simulation. While simulations are not as tightly controlled as scripted role plays, they are also not as free as improvisational role plays. Instead, they place participants in structured dilemmas with specific goals. This is a great forum for participants to practice their work scripts and acquire a comfort level with the new skill without the risk of the exercise becoming an improvisational role play. A simulation also allows participants to move around, an advantage for kinesthetic learners.

The most fun form of simulation is to create "secret fact" role plays, which involve participants getting together in pairs and each having secret facts that create opposing goals. This means that everyone in the room will have a chance to practice new skills, a requirement for adult learning. Each participant in the pair gets an information sheet that details who they are and what their situation is about. For example:

One participant may be an employee who is coming to his manager/ supervisor to make a harassment complaint, but wants to keep the conversation strictly off-the-record or he will not "officially" report it. He wins the role play if he can get the manager/supervisor to engage in an off-the-record conversation about harassment. Another participant is the manager/supervisor who knows she cannot keep conversations off-the-record and must report to HR any time she hears about harassing conduct. But she is very busy, late for the airport, and does not have time to deal with a report of harassment. The tension between the two participants' goals facilitates a role play for the practice of work scripts.

The different roles and experiences are debriefed by the trainer at the end of the role play. Remember, it is important to allow each person in the pair to get practice time. Thus, a fresh secret-fact scenario should be developed with new conflicts so that each participant has a chance to practice a work script too.

 **TIPS:**
- Secret-fact role plays are always well received because they are very interactive and do not involve being in a fishbowl, i.e., being watched by the group. Make sure everyone has a partner. If someone is the odd person out, then the trainer should buddy up.
- In order to make sure the secret-fact exercise is successful, the directions have to be very clear. For example, it is important to write at the top of every role play, "This is a role-playing exercise. You are to act out the part below with your partner. Do not read the script or information to your partner. Take time now to read your part and get into the role." It is also helpful for the trainer to actually demonstrate what will happen by using a volunteer and the materials.
- "Secret facts" is a great medium for setting up difficult situations that call for work scripts. Teach the work script first and then practice it with secret facts.
- **Done wrong:** Design a secret-fact scenario without opposing goals between the parties. For example, participant A wants to report unwelcome conduct and is willing to cooperate in all respects.

Participant B is delighted to take in the complaint and has the time to report it immediately. Ho hum . . . easy.

Games [Low-to-Medium-Risk Activity]

While there are numerous books for trainers on "games" that detail how to do specific interactive exercises, they do not all lend themselves to harassment prevention training because of the unique content issues and liability constraints. While it is important to keep the training interactive, it is also important to be cautious about using some games or exercises that ask participants to share their feelings about others based on protected categories such as race, religion, gender, national origin, etc. (See Chapter 7, "What Could Go Wrong?") When you choose games for your interactive training, keep those considerations in mind.

We have developed a unique interactive game called "Writing on Wallpaper" that is easy to adapt to any topic or customize to any culture. The Wallpaper exercise requires big easel pad paper. We use the kind that is a light plastic material that automatically sticks to almost any wall (without needing tape that could damage paint or wallpaper). There should be as many pieces of "big paper" as you have teams. You need the same number of permanent markers to write with as you have pieces of big paper, and one bell or whistle. For example, if you have four teams of six to eight participants you will need four sheets of big paper, four dark markers, and still one bell or whistle. Here is how it works:

Place or tape on the walls the separate sheets of big paper (wallpaper) and write on the paper such things as, "List below benefits of having everyone at our organization go through harassment training" or "List below how people may feel when they are disrespected." Such "List below" statements allow participants to think together as teams where all answers are welcomed and right.

If we have a group of 20 to 30 participants, we divide the group into four teams and put up four sheets of wallpaper with four separate "List below" statements or topics. Each group writes answers directly on the big paper. It is a lively exercise and we often pump music in the background. When we blow a whistle or ring a bell, teams rotate counterclockwise to the next wallpaper and make new contributions.

The trick is creating "List below" statements that will not elicit discriminatory remarks. It is important to demonstrate the first response as an example of the information you are looking for. The Writing on Wallpaper exercise is useful to get participants' "buy in" on various issues. Therefore, it is a great game to use at the beginning of the training. The responses should then be used by the trainer during general discussions to support the trainer's teaching points.

 TIPS: ■ Design "List below" statements that correlate to your legal teaching points and then incorporate them at the relevant part of your training. For example, "List below the benefits of taking prompt effective action to investigate all complaints." Wallpaper responses can then initiate your discussion about the employer's reporting procedures when you get to that piece of the content.

■ "List below" statements can also be used as an exercise on the tent name cards.

■ **Done wrong:** Designing "List below" statements that fail to anticipate illicit risky responses. For example, "What kinds of people are harassers?" While the response you may be looking for is "disrespectful people," it opens the door for participants to list protected categories such as race, gender, religion, national origin, sexual orientation, etc. Yikes!

Case Studies [Low-to-Medium-Risk Activity]

Case studies provide real examples of specific situations, problems, or events and are used to illustrate various teaching points. Trainers use case studies as a springboard for group discussions to highlight specific issues. Some ways to debrief case studies include asking specific follow-up questions or identifying good practices/bad practices. For example, a trainer who wants to use the case study method to teach the "reporting process" may tell the group the following real story:

> Mark, a manager, supervised Lee, who was up for a promotion. Lee sought out Mark to have an off-the-record conversation. Because they were such good friends Mark agreed to the off-the-record conversation and closed the door to his office. Lee confided that the president of the company had asked her out for a date knowing she was married. Lee told the president she was flattered, but she declined. Lee was never given the promotion. Lee asked Mark what she should do. Mark advised her that if she ever reported the president for harassment the HR department would never investigate the complaint because everyone could lose his or her job. Lee said she thought that would happen and asked Mark never to say anything about the situation. Mark agreed. A few weeks later the president instructed Mark to fire Lee without giving any reason.

While the above case study is based on a real situation, the trainer changed some of the identifying facts to protect the parties and the organization. Still it is rich with teaching points about reporting matters such as off-the record conversations, mandatory reporting requirements, prompt effective action, and retaliation.

Some discussion-based questions for the above case study include: "What were good practices the manager displayed?" "What were bad practices the manager displayed?" "Did the manager violate our harassment reporting process?" Additionally, you can ask follow-up questions that give participants the opportunity to apply the skills they have just learned about reporting. For example, "If you were the manager, what would you have done differently?"

 TIPS: ■ Case studies should be written down and handed out to teams along with the debriefing questions. Once the teams complete the exercise the leader can conduct a group discussion.

■ **Done wrong:** Ask participants if they have ever experienced anything like the conduct in the case study. Such personal questions violate the ground rules and beg for bigger problems like arguably putting the employer on notice of harassment issues.

Action Contracts [Low-Risk Activity]

At the end of the training, participants can complete "action contracts" or worksheets, which are specific commitments to apply to their daily work the skills they have learned in training. Because the participants decide what skills they will commit to using after the training is completed, they are more likely to follow through. Also, the mere act of reflecting on the training and writing out an action plan can further reinforce a commitment to change. For example, some action contracts in harassment prevention training prompt managers and supervisors to answer the following: "The one thing I will do to monitor the workplace for harassment is _____ ." Or, "Tomorrow I will walk and talk the harassment policy by _____ ."

 TIPS: ■ One unique way to handle action contracts is to have participants individually address an envelope to themselves and then place their action contract in it. One month later the trainer mails out all the letters and the participants receive their own gentle reminder to stay on track with their goals.

■ Remember, these are a handout, and therefore the "handout rule" about creating bad evidence applies.

■ **Done wrong:** Design action contracts that fail to have participants commit to a plan; or worse, ask managers and supervisors to answer in a way that implies that they have been acting improperly (e.g., "I will give up doing the following harassing things").

Methods of Training

Harassment training can be successfully delivered in many different formats:

Leader-Led

With leader-led training, a trainer facilitates the transfer of information to participants and gives them opportunities to acquire skills through practice. Sometimes this type of training is called "live training" because it involves a trainer who delivers a "live" interactive presentation to a group of participants. Leader-led training can also be accomplished via satellite, when the "live" location is being beamed to more remote locations. While the interaction in the remote areas may be less, there are methods to ensure involvement. For example, making sure the materials (tent name cards, pre-tests, simulations) are at the remote sites before the training begins is critical so that participants can interact with them. Also, designating a "captain" at each site to assist with the interaction, such as coordinating pairs of participants for simulations, is also helpful. The trainer should be interacting with participants in the live room as well as those at the remote locations.

Train-the-Trainer

Train-the-trainer courses are designed to teach trainers how to deliver a specific training workshop. For example, if an organization buys an anti-harassment training program for its HR staff to deliver, it will generally include a train-the-trainer workshop. In that workshop, the participants, who are trainers, will learn how to set up the room for the training, the content of the course, and how to facilitate the course, lead the exercises, collect documentation, and learn what equipment is needed. In turn, the trainer then directly delivers the course live to the employees.

Computer-Based Training

The technology of computer-based training (CBT) is ideal for learners who are self-motivated and can work independently on computers. While initial CBT programs were generally passive "read along" experiences, new technology allows for much more interaction by incorporating animation and video. Additionally, multimedia training programs are widely available through the Internet, on CD-ROMs, and via employer-sponsored internal web sites called Intranet sites. The advantages of such training lie in its ability to provide training faster to more locations, and usually more economically than traditional leader-led training. Still, there is some debate on whether it is possible for a computer to train a human on interpersonal skills. In any case, computer-based training is useful to reinforce skills learned in the leader-led training. To that extent, it can be considered a method to sustain skills.

The interactive approach, which is best used for adult learners, can be easily applied to computer-based training. For example, it is possible to offer a variety of interactive exercises, including simulations that give immediate feedback based on the participant's selected answers. The instructional designer's goal is maximum interactivity in order to enhance the learner's ability to develop skills. Other benefits of CBT training are the ability of the learner to go back to review materials not initially grasped, start and stop the training as desired, and move as quickly or slowly as needed.

In addition, CBT programs can offer excellent record-keeping services, including documenting the completion of the training. However, the technology should not keep a record of a participant's wrong answer choices, which may come back to haunt the employer in a legal dispute. Distribution and receipt of your harassment prevention policy can also be easily tracked. Many administrative platforms also keep track of which employees have completed the training and send e-mail reminders to those who need to start the training.

Self-Study Non-Computer-Based Training

Self-study is another method of delivering harassment prevention training to individuals, especially those who do not have access to a computer or cannot be part of a leader-led group, often because they are in remote locations. Self-study is also useful if there are only a few employees who need training.

Written self-study courses can be designed for self-starters and should include the same legal content that is in other courses. Self-study courses should also distribute your harassment prevention policy and a policy acknowledgment form for the participant's signature. One way to keep the written materials interactive is to have the participant complete written exercises as he or she reads along. For example, the participant could carry out a scavenger hunt on your policy.

New-Employee Orientation Videos

New-employee orientation videos provide an excellent way to introduce new employees to your harassment prevention policy with a focus on prohibited conduct, the reporting procedure, and retaliation. The video should be shown in conjunction with having the policy distributed and the acknowledgment form signed. While we do not consider such videos adequate as training, they are excellent tools to initially place new employees on the same page as others until they can receive training. New-employee orientation videos can also be used to share your policy with vendors who visit your workplace, as well as temporary workers and visitors.

LESSONS LEARNED

Dear Trainers:

Wow! I have dropped my flaming bowling pins in favor of communication cards, role plays, fun props, and wallpaper stations. The feedback is overwhelming that I have made a dry topic fun and interesting. I am so excited about my new style. Thanks for the great ideas. Any more tips you want to share?

Signed,

Ex–Fire Juggler

. . . .

Dear Ex–Fire Juggler:

Thrilled to hear you are trying new ways of engaging your audience. It really is a matter of trying new skills on and seeing if they fit. Here are the top 10 lessons we learned about delivery methods:

10. Design a course simply by mapping out the content you plan on providing and then selecting the interactive method you plan on delivering it by.[6]

9. As much as possible, shake hands with participants when they enter a room.

8. Engage participants from the moment they walk through the door with activities for them to do, such as tent name cards and pre-tests. Also, show them you completed the activities as a sign that you are participating too.

7. Start out all sessions with low-risk activities such as tent name cards, wallpaper stations, and Okay/Not Okay pre-tests (with obvious answers) so that participants feel safe interacting and learning.

6. Use different types of music to set a rhythm to the training and keep the beat going during breaks!

5. Use tent name cards to learn participants' names and personalize the experience—it makes for more engaged learners.

4. Use a variety of interactive methods during a training course such as pre-tests, video vignettes, role plays, hypotheticals, work scripts, and simulations to keep the training engaging, and just plain fun!

3. Create sets of standard communication cards so that participants can "talk" to you with them while they are listening to information, watching video vignettes, or engaged in role plays.

2. Always give participants the opportunity to practice the skills you are teaching or they may "learn" them but they will never "use" them.

1. Never present a lecture ever again if your goal is to help participants acquire knowledge and skills! Turn your lectern into a plant stand today!

Signed,

The Trainers

Notes

1. MEL SILBERMAN, ACTIVE TRAINING: A HANDBOOK OF TECHNIQUES, DESIGNS, CASE EXAMPLES AND TIPS, 3 (Jossey-Bass Pfeiffer 1998), *citing* H. R. POLLIO, WHAT PARTICIPANTS THINK ABOUT AND DO IN COLLEGE LECTURE CLASSES (Teaching Learning Issues No. 53, Learning Research Center, University of Tennessee, 1984).

2. SILBERMAN, *id.* at 3, *citing* D. W. JOHNSON, et al. ACTIVE LEARNING: COOPERATION IN THE COLLEGE CLASSROOM (Interaction Book Company 1991).

3. SILBERMAN, *id.* at 3, *citing* R. PIKE, CREATIVE TRAINING TECHNIQUES HANDBOOK (Lakewood 1994).

4. *See* ROBERT PIKE, CREATIVE TRAINING TECHNIQUES HANDBOOK (Lakewood 1994).

5. Use of music may require the purchase of a license for copyrighted works.

6. *See* Appendix H for a chart with a sample outline for a three-hour manager and supervisor program. Remember, you can use this to help you get started, but interactive training goes best when the trainer is comfortable and confident with the content, the audience, the interactive tools, and their risk level and timing.

6

No More Sweaty Palms:
Tips for Dynamic Delivery

Dear Trainers:

One last concern before I take off to do the training. I think I understand the content well enough to do harassment training, and I have already "translated" the teaching points into role plays, simulations, and hypotheticals. I even made tent name cards and an engaging True/False pre-test to get the training started with pizzazz. All those things now are in my briefcase, but when I went to pick it up, my hand slipped off the handle because my palms were getting sweaty. I wonder if I can handle large groups where some participants may give me a hard time, while others may be too embarrassed by the topic and not participate. Can you make it "sweat proof" for me?

Signed,

No More Sweaty Palms

. . . .

Dear No More Sweaty Palms:

Every trainer has feared a tough crowd. The reality is that you will probably have one at some point in your training initiative. And, if your experience is anything like ours, you will usually meet such a crowd more than once. But consider these crowds a "gift" to you because they give you an opportunity to hone your "stand-up" training skills and make you a more dynamic, versatile trainer.

Folks who don't like the messages you are delivering may verbally resist them in an overtly argumentative, hostile manner, while others may refrain from participating even if you stand on your head while playing a harp with your toes. And then, of course, let's not forget the "dozzzzers" who want to sleep through it because they were sent to you by others and don't think the training applies to them. Don't worry, though! There are many simple techniques that we have successfully used to handle all of these participants. You will learn how to keep them engaged without losing their dignity (or yours). Read on!

Signed,

The Trainers

Overview

In no other self-respecting job can one say "sex" as often as you will, and get paid for it! Interactive harassment training requires its trainers to be not only comfortable with the content and dynamic in their delivery, but also in complete control of the training so that the participants stay engaged and the training stays on task.

How you are perceived by the participants can persuade them to listen and interact or dissuade them from doing so. Thus, being aware of how you present yourself through your words, actions, clothes, body language, and voice are all critical to the success of your training. Perceptions are formed by receiving input, processing it, and then making judgments on it in context with our other experiences. Compelling trainers understand the power of perceptions and deliberately create themselves and their training accordingly.

Let's look closely at the multiple tensions a trainer must successfully balance during this unique training: (1) observing legal accuracy, (2) managing resistance, and (3) achieving active participant involvement. Interactive harassment training is about critical, real, legal issues and concepts; everything said in the training has legal importance attached to it. So it is important not to say anything that is legally inaccurate. For example, "The laws protect only women so only women can sue." Wrong! Another tension is that harassment is often a divisive and controversial topic. And as with any controversial topic, it may be met with resistance or hostility—in other words, tough crowds. The third tension has to do with maintaining the interactive training itself. This is not the type of training where a trainer can hide behind note cards or PowerPoint slides, but a training where delivery skills take on great importance. For interactive training requires the audience to become actively involved in the discussion, and, to a certain extent, assist the trainer in teaching concepts and then rehearsing the skills in a safe, nonjudgmental environment. The challenge for the trainer becomes how to keep the audience actively involved in a controversial, legal topic, while still maintaining control and facilitating change. The platform skills necessary to meet this challenge are the focus of this chapter.

What You Need to Know . . .

Why the Room Setup Impacts Your Effectiveness

Imagine how effective you would be presenting to participants' backs. We have attended trainings where chairs have been set up that way around round tables. Half the group faced forward while half faced backward. While this seems like a simple issue, it goes unrecognized by trainers who solely focus on content or who are not involved in the logistics of setting up the room.

The training room should be set up with either round tables and chairs positioned in a half-circle facing the front or in a "U" shape. These setups allow the participants to feel they are part of a small team with round tables or part of a large group with the "U" setup. Does it matter which way to set up the room? Yes, sometimes. If you know groups will not be resistant to the training, they will do well in the round table setup. More-resistant groups are easier for the trainer to control in the "U" shaped setting. Equipment such as easels, TVs, and flip charts should all be at the front of the room facing the audience.

Skip the tall podiums or the stage. It separates the audience from you, especially for smaller audiences (under fifty participants). Podiums and stages are better used for ballroom presentations and formal speeches.

It is most helpful to have your own "trainer's table" to place your agenda, role plays, case studies, and props on. Set things up in the order that you will be using them. Your table should be set up near the front and to the side to make it useful for you. You can also put your watch on the table to subtly keep track of time. Have a box nearby so that when you are done with an item you can put it in the box and keep the "trainer's table" free from unneeded clutter.

 TIPS:
- Check all seats to see if all participants can see the front of the room and the TV and read writing on the easels. This will help you to manage your volume control and font size.
- Keep the presenter's table neat so that you are perceived as being organized.
- If you are giving every participant props such as a set of communication cards, a pre-test, and a tent name card, put a set of the same props on your presenter's table to remind you to use them.

Otherwise you may forget to use them and at the end of the training folks will come up to you asking what they were for.

■ Always have the setup done before participants are likely to arrive. You don't want to be the host of the dinner party who is still setting the table when the guests arrive!

■ Having told you what is ideal, let's talk reality. Sometimes you will have to train in imperfect circumstances and environments. We have, many times. Do the best you can to create an environment where participants can see you, hear you, and feel comfortable, and you will rise above the environment.

Before the Training Begins

Why wait to handle resistance until it rears its ugly head in the middle of your training? There is only one way to keep it out of the room, and that is by planning for it before the first participant walks through the door. Think of yourself as the host of an event. Yes, harassment training is your event and, as with social functions you may have hosted, you have a definite role. No guest feels welcomed if you are still running around setting up handouts, checking out the equipment, or reviewing your notes. Rather, the stage should be set up in advance so that you can meet and greet your guests at the door. This is the first step in reducing resistance, establishing a personal experience for the participants, and creating buy-in to the messages that will be appearing on the runway in your act. Have the music going, and as participants walk through the door, stick your hand out for a friendly handshake, welcome them to your training, introduce yourself by your first name, ask their name, and encourage them to sit anywhere they would like. As soon as they get settled, show them the activities you want them to get right into.

Along with creating interactive exchanges, what this also accomplishes is to eliminate the "spotlight moment." Perhaps the most difficult aspect of training or public speaking is that first moment when the speaker walks in from the wings and all eyes follow him or her to the front of the room, waiting for pearls of wisdom to flow effortlessly from the speaker's mouth. This terrifying moment will be all but eliminated if the trainer has engaged in light banter and "meet and greet" conversation with the participants beforehand. You will have gotten your "voice," and the training will simply be a continuation of the previous conversations. You'll see, this simple but critical step to your training does several things: it helps participants to like you, which makes it more likely they will "buy" your message. In addition, you will feel more comfortable when the training actually begins.

TIPS:

■ Have a master checklist of materials and equipment you will need for every training so that you are organized and don't have to reinvent the wheel every time you train on harassment prevention.[1] There are some items you will have to make sure everyone receives, while other items are only one per class. For example, everyone should receive a tent name card, pre-test, a copy of the policy, a policy acknowledgment form, one set of communication cards, and an evaluation form. If you use tools like role plays, case studies, and Wallpaper, you will need to figure out how many copies and sheets you need of each. If you use video vignettes, you will need a TV and a VCR. And never forget the music and boom box or refreshments.

■ The "meet and greet" method allows you to meet the participants one at a time as they filter into the room. Remember, some people like to come earlier to meetings—you don't want them watching you set up and you certainly don't want to make them wait outside in the hall when they could be sitting down and being productive doing something else. So get there early!

■ If possible, every participant should get the "meet and greet" experience. Never underestimate the value of shaking everyone's hand—it comes in handy when you deliver the tough messages.

■ Meet and greet folks with your tent name card and your pre-test completed so that participants know you are not asking them to do something you haven't done. Heck, if you can do it, how hard can it be? Tease them that they can look at your answers and, if they need help with their name, can ask their neighbor. Humor (the nondiscriminatory kind) helps build camaraderie and reduce resistance to the training.

■ Schmooze. Yes, walk around the room and make sure participants are comfortable when you are not at the door meeting and greeting. Ask them if they can see the front of the training room, see the TV, are their markers working, are the chairs okay? Point to the coffee and cookies if you have refreshments and encourage them to enjoy. This is a great time to ask people what they do at the organization (if you don't already know), how they found out about the organization, etc. Show you are sincerely interested in them and want to get to know them professionally, and you can even reveal information about yourself using these safe questions. The operative word is "sincerely," so listen with your ears and eyes to their answers. If you do this right, they will find it difficult to give you a hard time later. It's your party and you'll schmooze if you want to.

■ **Done wrong:** Sit at the front of the meeting room or stand at a podium reading your notes or the newspaper until the time of the training begins. Then try making personal contact with thirty-five people all at once during your opening introduction. Good luck!

Your Introduction: Creating Your Persona and Setting Up the Experience

Give me a break, isn't just saying "Hello my name is and my title is and today we are going to discuss harassment" enough? No. Your one-to-two-minute introduction is the most critical personal moment between you and the participants. It will help them to like you and trust you, and will increase retention of your messages while reducing potential resistance. To take it lightly and fail to prepare an interactive credibility-based introduction is a big mistake.

Control will be more easily lost if the participants feel the trainer is not comfortable and confident. Thus, the introduction must establish the trainer's mastery of the subject and confidence in delivering it. For it is at the introduction where the trainer is being scrutinized and assessed by the participants. Participants will make their initial judgment as to whether they should believe the trainer, argue with the trainer, be persuaded by the trainer, or, most important, learn from the trainer. One way to maintain this control and to establish credibility is to know your personal introduction cold. The delivery must be smooth. There should be no hesitation. No notes. No nothing—except the trainer engaging the participants with nothing in between.

In your introduction do you want the audience to know what your academic background is, your school grades, that you made a papier-mâché mouse in sixth grade that won first prize? No. You want them to connect to you through your introduction. It is the critical time to share with the audience your training persona, or personal introduction—"the professional expert who is human, caring, and fun." Designing a well-developed training introduction will help you to further cement your rapport with the audience.

The goal of the trainer's persona is twofold: (1) to build credibility, and (2) to make yourself human.

Build Credibility

With respect to credibility, the trainer must immediately answer the question, "Why are *you* standing up there?" So there must be some mention of your credentials. For instance, if you are a human resources person, you may mention that you have investigated numerous types of harassment claims or if you are an attorney you may note that you have litigated harassment claims in federal and state courts. Really, the goal is to communicate expertise without appearing to be pompous. Unless you have expertise in the matter, there is really no basis for you to be leading participants in the class.

Make Yourself Human

By making yourself human, we simply mean that you also should reveal something about yourself relevant to the training. This should not be overly personal (your marital status) or irrelevant (you like to swim) but should give the participants a little insight into who you are; that you are not simply a drone fired off by management, but an engaging, interesting person. We highly recommend humorous anecdotes about yourself as one way to make yourself human to the class.

For example, one trainer created the persona of "bad-doer turned good-doer" and introduces herself by saying,

> I am delighted to be here because many years ago at my prior company, I was the poster child for harassing conduct. I was the queen of "e-dirty jokes." Everyone used to laugh at my daily round of sex and ethnic jokes I pulled off the Internet while at work. I thought I was the most popular employee on the cubicle circuit. However, after going through training many years ago I realized how I probably offended everyone I worked with and how I probably should have been fired for telling the jokes and misusing the Internet. Now I work here and spend most of my time conducting harassment prevention training so that people like the old me become extinct and we have more respectful workplaces in which we can be productive.

This type of story, done well, makes the audience feel that the trainer is human and has an intrinsically good purpose for being in front of the room. In contrast, an opening like, "I went to Harvard Law School and had the highest grade point average . . ." will flip off the participant's receptivity switch and flip on the resistance switch.

 TIPS: ■ You can start your introduction with a question that will get every hand in the room to shoot up. For example, based on the above introduction the trainer could ask, "How many of you have ever laughed at a joke even if you didn't think it was funny? Raise your hands." This technique creates interactivity even during your opening and gets the audience involved. It also sets the foundation for you to leap into your human story.

■ Never tell a personal story about criminal activities, sexual prowess, or the like. Your goal is to humanize yourself, not to discredit yourself.

■ Never start a presentation with a joke or story unless you are absolutely comfortable doing so and the joke or story is appropriate. Unless you are a trained stand-up comic, it is the fastest way to bomb. One more word along these lines: Don't ever try to be someone else. Be yourself. Participants can spot an "impostor" in the first two minutes!

■ **Done wrong:** "Hello, please sit down and be quiet. This is going to be a very important training program and I am the one delivering it because of my vast knowledge of the law and because I am a role model of good conduct." Need an antacid?

Once the persona has been established, the next question that you must answer is the #1 question for the participant, "Why do I need to be here?" or "What's in it for me?" or perhaps—to put it most bluntly and accurately—"Why should I stay awake?" To answer these questions, the trainer must know who the audience is.

Why Am I Here?

As discussed previously, a trainer must gear his/her presentation to the particular audience. It is at the introduction where this must be addressed directly. The "Why do I need to be I here?" answer varies depending on the audience. To upper-level executives, the trainer's focus could be on money, the bottom line, how much defending and potentially losing a harassment suit can cost the organization. There is no shortage of cases and statistics that will strike fear in the heart of every able-bodied executive.[2] With managers and supervisors, the emphasis should be on responsibility. In no uncertain terms, the message should be that their job depends on enforcing the anti-harassment policy properly; that they are the gatekeepers of the policy and responsibility for an environment of disrespect and harassment will fall squarely on their shoulders. Finally, the rank and file must receive two messages—first, that failure to follow the policy will result in discipline, up to and including termination. Simply put, their job hinges on understanding and abiding by the policy. And second, if any one of them is experiencing harassing or retaliatory conduct, there is a process they can use to report it.

TIPS: ■ If you can, in advance of the training, ask others in similar positions to those of your participants why they would want to know about harassment. That way you are not guessing when you answer the question, "Why do I need to be here?"

■ Boil down your knowledge of the audience and "Why do I need to be here?" to the most personal level you can. In other words, it is more compelling to talk about how Mr. or Ms. Participant will personally be impacted by the training than to talk in abstract terms.

Giving the Road Map, Agenda, Goals, Housekeeping Issues

Participants are more comfortable and participate better when they know where you are going to take them in your training, from the beginning to the end. Therefore, it is wise to take a few moments at the beginning to tell your audience the agenda or "road map" for the session, the goals of the training, and various housekeeping issues. This means you should share the topics you will cover, the policies that will be reviewed, the skills they will acquire, the documentation they will sign, the participation that will take place, and the breaks that are planned. Setting the audience's expectations up front allows the audience to meet them.

 TIPS: ■ This information should be shared with enthusiasm to send the subtle message that this will be fun and interesting.
■ **Done wrong:** Save time and never tell your audience about the journey they will be going on. Watch the audience wriggle and squirm wondering "What's next?" and "When will this really end?" The suspense should kill . . . the interaction.

The Ground Rules

Ground rules can be a critical part of setting up a safe training. (See Chapter 7 for some issues that can be handled as ground rules.) Ground rules are also a good resource for handling difficult participants (see below, Managing Difficult Participants). Therefore, they must be shared with the audience after your introduction and the road map issues have been addressed. Review them in an upbeat but firm and friendly voice and then move on.

For example, one ground rule list could read something like this:

- We are not going to debate the laws, court decisions, or the policy.
- I am not here to give legal advice.
- We cannot discuss any current or pending situation.
- No discriminatory remarks during the course of the training, or ever.
- We cannot determine what is or is not harassment. Only a judge or a jury can draw that conclusion. Instead, we must focus on what is or is not a violation of the policy.

TIPS: ■ When reviewing the ground rules let the audience know they will be easy to follow because they are all about time management issues (no debating the laws or policies) and respect for one another (no discriminatory remarks).

Debriefing Tent Name Cards

As described in Chapter 5, tent name cards serve many critical purposes to facilitate a successful beginning to the training experience. They will help you learn your participants' names and help them remember the ground rules that you have printed there. Additionally, debriefing the tent name cards as soon as possible in the training lets the participants have their "voices" heard early to get them invested and engaged in the experience. When folks participate in a low-risk exercise soon after the training begins, they are more willing to take the next step and get further involved in the experience. Therefore, if groups are sitting at round tables you should have them take a few minutes to share with each other their names and responses to the exercise. If they are sitting in a "U" formation, make groups of three to four participants and have them debrief the tent name cards. Don't forget the music and remember to schmooze from table to table listening in, chatting and building personal relationships. At each table appoint someone to serve as captain to share information about their team with the whole group. (For example: appoint the person wearing the most blue or who lives closest to the facility.)

TIPS: ■ The platform skills needed for this exercise are similar to those for a game show host, giving every team a chance to be introduced and share some safe information. Be interested, listen, and thank them for sharing.

■ This exercise gives you a few more minutes to learn something about the participants. Your genuine concern will help further reduce their resistance and build retention for your upcoming messages. Listening builds goodwill.

■ **Done wrong:** While teams are sharing their information with the group, you are looking at your notes for the next activity. If you looked up, you would see the bubbles over their heads read, "You don't care about us so we're going to give you a run for your money!"

Delivering Interactive Training Exercises

Keep your "host hat" on while engaging the audience in various interactive training exercises. You will find you utilize the same social skills you needed to meet and greet the participants at the beginning of the training. Be friendly and helpful. Walk around the room and make sure all the tables understand what to do during the group exercises. Offer useful ideas or tips to groups while they are debriefing answers. Show them you care and want them to get the answers right. For example, if you have designed a policy scavenger hunt exercise (see Chapter 5) to debrief the policy, then walk around and talk to all the teams and see if they are getting the answers right. Show them where the answers are in the policy and praise them for working so fast and hard. Use humor to engage participants. You are there to inspire activity and learning in a fun environment, not withhold answers and whack knuckles.

 TIPS:

- Make sure to demonstrate the results you are looking for. For example, if you want participants to use their Okay/Not Okay communication cards while they are watching a video vignette or role play, then use yours too.
- Practice giving clear directions. We have found over the years that even simple tasks call for multiple-step directions. Draft instructions in a step-by-step format on an index card that you keep on your trainer's table.
- **Done wrong:** Sit at the front of the room putting dates in your calendar while participants are working on exercises. Better yet, step out of the room to make (or take) a call on your cell phone. You just might see participants passing you by in the hall to do the same!

Managing Difficult Participants

Consider difficult participants a gift to you. Yes, another gift! They come in all forms, from those that want to challenge your expertise and argue every point including the spelling of your name to those who answer sarcastically. But with a little strategic planning, he or she will allow you to show the other group members that the training room is a safe place to learn and you are in complete control.

Beware! Difficult participants may cause the trainer to lose control of the class. A mini-battle is formed between the disrupter and trainer as to who will gain the ears of the rest of the class. It goes without saying that if the trainer loses this battle, the training session itself will be less than fruitful. Not only does the loss of control probably result in the loss of participant attention, but there always is a risk that the disrupter may make statements that are at odds with the trainer's message and, in fact, may create liability. If the audience is paying more attention to the disrupter, the audience may remember what the disrupter says as opposed to the trainer, and one can only *guess* what the disrupter will say. In addition, participants who find the disrupter obnoxious will blame you for not keeping control of the class. That is why it is so important for trainers to anticipate and prepare for difficult participants. Just as a trainer should be prepared and comfortable with the harassment materials, so too must the trainer be prepared and comfortable in dealing with a difficult participant in order to keep control of the class.

When a difficult participant either questions or makes comments about nearly everything said by the trainer or the class, the trainer is faced with what we call the "resister." The resister, the most common of training enemies, has an opinion on everything, and the opinion usually differs from that of the trainer or the other participants. The resister also has no qualms about verbally sharing his opinion. So how should a trainer handle a resister? There are several methods.

In general, when dealing with either questions or comments from any participant, a trainer must make every effort to respond and validate in a positive way the offering from the participant. A trainer must never ignore a question or a comment. Ignoring a question or a comment from a resister not only results in the resister gaining more control of the class, but may legitimize the comment or question through the trainer's omission. So the question or comment must always be responded to. But how?

The first method we recommend is called the "headline answer technique." This technique consists of validating the question and then answering the question with a quick sound bite or abbreviated headline. What we mean by validate the question is that a participant should never be made to feel that he or she made a stupid comment or asked a stupid question, no matter how stupid or ridiculous the question or comment may actually be. For example, if a participant asks, "If I want to report harassment, but my immediate supervisor is on vacation for a week, do I have to wait for her to come back?" Headline answer: "No. You can go to anyone else in the reporting procedure." Validation: "Great question, thank you for asking. You will learn more about that when we cover the reporting procedures under the policy." Headline answers should satisfy such inquiries without frustrating the participant with an answer like, "You will get that answer in another hour and a half."

On the other hand, if the question or statement looks like it is going to violate a ground rule or take the group into a big black hole of debate, then immediately go to what we call the "double thank-you technique." First, thank the resister for their interesting observation, and then thank the resister again for reminding you about one of the specific ground rules.

For example, occasionally a resister starts the song and dance that "the policy won't let us do anything and I think it is a crock of . . ." If this happens, you must politely cut the resister off by saying, "Thank you for your interesting observation. And thank you for reminding me about the ground rule 'no debating the policy' because while I understand that people could have many different perspectives about the policy, we just don't have time in this short period of training to get into those discussions or the whole group will have to stay later than planned in order to get through the material."

This is the best way to shut down a resister, help the resister save face by recognizing the value of his/her comment, and avoid a *West Side Story* gang war. In fact, the group will support you because no one will want to stay later than necessary to satisfy one or two hostile individuals in the group. This method is almost foolproof and works for any type of resister. All you need to do is pick the right ground rule to thank the individual for. And don't forget, you can always invite the individual to talk to you about his/her observation after the training is over. They rarely do because they do not want to waste their time. They want to waste yours.

We have used many different methods to handle resisters, but we recommend these two because they work the best under most circumstances. Even in difficult classes, when trainers use these techniques, they receive very positive comments like, "Handled difficult situations very professionally" and "Listened to everyone politely." As we always say, the proof of whether or not a technique works is . . . in the evaluations.

 TIPS:

- ■ Never get angry with a resister. Losing your cool will cause you to lose your credibility with everyone. It will also chill the interactivity level you still need from the other participants.
- ■ Keep good eye contact with the resister. Walk over to where he or she is seated and deliver your techniques up close and personal. But do not stand over the participant and be intimidating, as that will only escalate things.
- ■ If these techniques don't solve the problems, it may be warranted to pull the resister aside at a break and give him or her a more constructive role. Of course, the final move, and only as a last resort, is to ask the individual to leave the class.
- ■ **Done wrong:** "Be quiet and knock it off. You are wasting my time!" You might as well have said, "Now, everyone else turn on me at the count of three!!!"

Managing Quiet Participants

Some participants are quiet and don't like to interact in a group or in front of others. Respect these folks as long as they are engaged with you in eye contact and interactive exercises. People learn in different ways and at different paces. As long as your training is designed to be interactive and the quiet participant is interacting, then learning is probably taking place.

 TIPS:

- Don't force quiet participants to take on "spotlight" roles such as role plays.
- If your training is designed with plenty of small group activity, the quiet participant may be participating there.
- Recognize and praise everyone's participation, especially that of the quiet participant, to increase participation by everyone.
- When doing small group exercises, you can approach the quiet individual and ask for his or her thoughts on the issues at hand.
- **Done wrong:** Ask the quiet participant, in front of the group, why he or she is not participating and then assign that individual to debrief an exercise in front of the group.

Handling Questions

You must always be ready to handle difficult questions and those you do not know the answer to. As a reminder, you are human and therefore it is okay not to know the answer to every question that will be raised in all your future training classes. The successful way to handle moments when you do not know the answer is to go back to the group, asking, "Does anyone here know the answer to that question?" Or, "Does anyone in the group have any experience with that?" If no one does, or you do not feel participants are the proper place to get the answer, you can always say, "I don't know, but I will check it out and get back to you." Just make sure you do.

 TIPS:

- You may also consider the "toss-back" technique, where you say, "Great question. I do not have an answer for you, but based on your experiences what are your ideas on that?"
- **Done wrong:** "Wow, you stumped me on that one! I haven't got a clue. If you find out will you let me know?" You never should give up your credibility.

Handling the Game of "Question Bombardment"

Often a participant will pepper the trainer with a multitude of questions. Regardless of whether the questions are good or inane, it becomes apparent that the goal of the participant is to hear himself or herself talk. Not only do an exorbitant number of questions throw off a trainer's schedule, but constantly hearing one voice from the class will often create a boring or annoying disruption for the rest of the class.

As with a resister, the trainer must validate, answer, and move on. But at some point, the trainer must stop the constant flow of questions. Unlike a resister, who is generally antagonistic, a person who asks too many questions is often sincere and not intentionally trying to upset or upstage the class. So the trainer should deal with the questioner in a nonconfrontational manner and with more humor. For example, the trainer may tell the questioner that his/her quota of questions has been used up, or that the questions have been great, but someone else should be allowed to participate. As a last resort, the trainer may want to turn the tables on the questioner and have the questioner try to answer his/her own question.

End on a High Note

Some trainers never know when to end, so the audience will send subtle clues like slamming notebooks closed, looking at watches, and getting up to stretch. Not so subtle sometimes. You should pay as much attention to the end of your presentation as to all the other parts. End on a high note that gives the audience a sense of hope in the brightness of the future. For example, recap all the teaching points by reviewing the pre-test and telling the participants that they now have all the skills they need to successfully follow the organization's anti-harassment policy.

 TIPS:
- Remember to shake hands with participants on the way out and thank them for their invaluable participation.
- **Done wrong:** Abruptly announce during an exercise, "Oh my gosh. We have run out of time. Thank you and good luck!" You will need the luck when they abruptly complete your evaluation! All training sessions need a sense of orderly closure.

LESSONS LEARNED

Dear No More Sweaty Palms:

We have put together some lessons learned on presentation skills:

The "Never" List
- Never read your presentation. Yawn.
- Never look at your watch.
- Never jingle coins and keys in your pocket.
- Never use profanity.

- Never look at people's hair or foreheads rather than their eyes.
- Never fail to give scheduled breaks.
- Never lecture or preach.
- Never keep your cell phone or pager on.
- Never point. It's rude. If you use your hands to express yourself, keep your fingers together and palms up.
- Never have odd mannerisms like twisting your hair, playing with your glasses, or picking the skin off your lips. Yuck!
- Never overdress or underdress. Wearing a suit to a manufacturing plant when training hourly employees is like wearing a tuxedo bowling. Wearing shorts and open-toed sandals to a training is like wearing a bathing suit to a shopping mall. You should be properly dressed, one step above the audience.
- Never complain or be critical.
- Never use audiovisual equipment you don't know how to turn on or operate.
- Never sit.
- Never wear miniskirts, dangling earrings, or sandals. Nix charm bracelets and holiday jewelry pins that light up.
- Never give handouts before you are done explaining what the audience will do with them. Otherwise, they get distracted with passing them around and looking at them, which means you will have to give the instructions a few more times.
- Never pace. The eye follows movement. While some movement is necessary to talk to folks in the room, unnecessary pacing and shifting can make the audience seasick.
- Never show nude photos, tell sexual jokes, etc., as examples of what is prohibited conduct.
- Never have your working table be a mess with your notes, coffee, newspaper, props, pens, markers, briefcase, etc. Keep it organized-looking and try to keep the things you are not using, like your briefcase, out of sight. Looking organized gives you credibility.
- Never use hostile body language like closed arms, crossed feet, hands in pockets, judgmental grimaces, or discouraging frowns. Shaking your head in the "no" motion while someone is participating will chill all others from trying.
- Never say "okay" when you mean to praise an answer with "good answer," "that's right," or "you hit the nail on the head." "Okay" is not "okay" as a positive response.

The "Always" List
- Always start the training on time and end on time. Respect people's other commitments.
- Always be gracious. Thank those who helped you set up the training or invited you to do the training, even if they are not in the room.
- Always have participants sit up front if the room is half full. It makes for a more intimate experience. Latecomers can sit in the back without disrupting the group.

- Always have a glass of water nearby for dry mouth, the trainer's most common ailment.
- Always discuss the topic from the insider's perspective. Use "us and our" over "you and yours" to connect better to the audience. For example, "Our policy . . ." creates a better connection to your audience. In contrast, "Your policy . . ." sounds more like an outsider telling those of us on the inside what to do.
- Always carry an extra extension cord if you use equipment.
- Always show the noun when you talk about a noun. For example, if you tell the participants to take out their pre-test, hold yours up to show them what you are talking about.
- Always use your voice effectively. Monotone voices lose listeners. So vary your pitch, volume, pace, tone, and pauses to keep your presentation . . . [pause] interesting.
- Always practice your full presentation before your first training. Consider it a dress rehearsal by using the equipment, props, and tools such as role plays.
- Always wear comfortable shoes. Four hours of standing on your feet feels more strenuous than running nine miles.
- Always try to learn the names of the participants and use them. Move tent name cards around so that you can easily read them.
- Always act enthusiastic and listen with the same level of enthusiasm.
- Always know the quirky room issues you are training in before you start. Once we turned off all the lights to test the TV/VCR to see if we could see the video vignettes from all angles of the room. To our horror, it took twenty minutes for the lights to warm up in the room again. We showed the videos with the lights on and explained why.
- Always smile and nod in agreement when people are interacting with you with the right answers. Everyone needs to know when they are on the right track, even groups.
- Always know whom to contact for assistance with problems in the room. Too hot, too cold, too bad if you don't know whom to call.
- Always maintain eye contact. If you have a big group, maintain eye contact by addressing participants like a clock, at the points of 12, 3, 6, and 9.
- Always offer candy and refreshments. Everyone is perceived to be a better trainer when he or she hands out candy to all the participants during exercises or has refreshments in the room.
- Always be culturally sensitive. If a participant writes "Mr. Lee" on his tent name card, then call that person Mr. Lee and don't demand to call him by his first name.

Signed,

The Trainers

Notes

1. *See* Appendix I.
2. *See* Appendix J.

7

What Could Go Wrong? Making Your Training Disaster-Proof

Dear Trainers:

I have a handle on the content, the way the course is put together, and my delivery. Life is good!

Signed,

What Could Go Wrong?

. . . .

Dear What Could Go Wrong:

Don't even ask! Almost anything!! From resistant participants who want to leave the room to lost documentation to taking your training to trial—and defending it—lots of things can and do go wrong.

Savvy employers are realizing more and more that the best way to protect themselves against lawsuits and financial liability is through effective policies and training. Of course, as the number of employers conducting training increases, so too does the number of employers who point to training for protection during subsequent lawsuits.[1] The logical end result is that training has become and will continue to be an instrumental battle in the litigation wars. The bottom-line question is: "Can you take your training to trial?"

Because despite all best efforts in training or elsewhere, an employer may get sued. This is reality. Since training can be a key component of an employer's defense, it is important to explore training from the other side. How might your training be attacked in litigation—at summary judgment or at trial? How might your training fail to meet the affirmative defense? Let's look at the pitfalls with the benefit of our hindsight and experience.

Signed,

The Trainers

Overview

The focus of this chapter is to consider how training *can* fail—so that it *doesn't*. Certainly, the first thing to examine when auditing a training program is the content and design of the training. What's in your training, and how you design and present it, will have an impact on whether you can successfully take your training to trial (as well as whether you are successful in delivering the harassment-free workplace message). We are confident that a training program that mirrors our chapters on content and design will not only help an employer meet its affirmative defense, but go a long way in limiting liability. Yet even as we pat ourselves on the back, we know from our own painful experiences, as well as those of others, that knowing what content and interactive tools should be in a program, and designing and delivering it well, does not necessarily mean that the program will not be challenged. Or go well. So along with our admonition to carefully audit one's policy and training against our chapters on content and design, we invite you to ask yourself (and answer) the following questions.

What You Need to Know . . .

Have You Trained from Top to Bottom?

Was everyone in the workforce trained, or were certain people excluded? Often training does not encompass certain groups of employees. Many employers exclude some members of their workforce for what they claim are "practical reasons." The most likely suspects to be excluded are employees who work outside of a main facility (for example, salespeople) or upper-level executives (who are often "too busy"). There are a number of reasons why this issue is so significant and why your training should be mandatory for *all* employees. The most obvious reason is that if the employee who committed the harassment or was harassed did not go through the training, the training may be useless in establishing the affirmative defense.

Another reason this is so problematic is the resonating message that is sent when you fail to train everyone in the workplace. The message? That the employer is not totally committed to anti-harassment in the workplace. This may or may not be true, but hypocrisy rings loud and clear when upper-level executives are not being trained and all other employees are going to training. Employees may hear from the trainer that the anti-harassment policy is important, but the message will be lost if the leadership is not trained. Certainly, if you take your training to trial you will be asked if you trained everyone. If you did not, you can expect to have that decision attacked.

Not only must upper-level employees be trained, but all training sessions should consider having welcoming words from a leader of the organization. Whether it is live or from a video or a written message, it is imperative that all employees feel that the training is supported and followed by everyone in the organization.[2]

Another reason to train upper-level employees is that such employees may be viewed as an alter ego of the employer. Which simply means that if an employee is of a high enough rank in the organization, the employer is strictly liable/accountable for the employee's actions, even if the actions violate organization policy, or even if the employer arguably can assert an affirmative defense.[3] Your only hope here is to make sure that your high-ranking people know absolutely what they can and cannot do under your policy.

 TIPS: ■ In some organizations where we train, upper-level executives go to training along with everyone else. This is absolutely fine in some cultures and it does send the message that leadership is committed. In other organizations, however, a special session is arranged for a "leadership team." This allows for a discussion of the entire training initiative with the executive team and a reminder of its importance to the organization. (See Chapter 3 and discussion on content appropriate for executives.)

Is Your Content Accurate?

This seems to be an obvious parameter of any training. But if we had a nickel for every time. . . .

Don't believe us? How about where the person responsible for the sexual harassment training testified at her deposition that she believed that a male supervisor would not be engaged in sexual harassment if he simply apologized after exposing his genitalia.[4] We use this example for a couple of reasons. First of all, this wasn't a rank-and-file employee misinterpreting or misunderstanding the training, but an upper-level manager who had responsibility for the training itself. What should that tell you? That accuracy of the content is of the utmost importance because if you get it wrong it will certainly show up at trial. Disseminating wrong information may be the first place of attack for a plaintiff. Besides, you don't want to give inaccurate information to your participants.

Associated with accurate content is ensuring that the tone of the training is geared toward the right audience. Remember, content will vary depending on the particular audience. (See Chapter 3.) One mistake that trainers sometimes make is to create one program for all audiences, rather than customize based on what the audience needs to know. The risk, of course, is providing information that is not needed or omitting information that is needed.

TIPS: ■ We see a lot of training programs that focus on high monetary verdicts in all of their sessions. While we highly recommend that upper-level executives receive information on trends in cases and high monetary verdicts, providing such information to the rank and file may provide employees with an incentive or "road map" to sue. It also makes legitimate plaintiffs seem like "gold diggers," which is not accurate.

■ Also, failing to discuss quid pro conduct at all may be considered playing hide-the-ball. We have looked at some training programs that refuse to discuss this topic with their employees for fear it is happening to them. Willful blindness does not absolve an employee from liability. While it may not need to be discussed in such detail with employees as it does with managers, it should be addressed as conduct prohibited under the policy.

Are Your Reporting Procedures Accurate?

Here is a problem that we have encountered on numerous occasions and with many clients of all sizes. It is a problem that, if not corrected, will badly damage your chances of establishing an affirmative defense. Are the reporting procedures that you are training on accurate and up-to-date? Before doing training, an organization must take a long, hard look at its reporting procedures.

More specifically, are the telephone numbers given to report harassment accurate and effective? Are the titles of the positions correct? Are there multiple avenues an employee can use to complain? And finally, do those people, whether they are managers, supervisors, or Human Resources people, know that they are designees for reporting, and are they competent in dealing with complaints?[5]

TIPS:
- When we do training for a client, we call the hotline number(s) to try to make a report and see what happens. It is unbelievable the number of times we either can't get through, or the line is busy, or the line is disconnected, or no one ever returned the call. Training on a hotline number that does not work is "bad news" for your affirmative defense—and your organization!
- Look at your policy. List the job titles of the people who are authorized to accept complaints. Ask yourself: Have they all been trained? Hopefully, the answer is yes. If it isn't, get busy!
- After you do this, ask yourself one more question: Are there any other job titles not listed in the policy that might be prime candidates for reporting ("leads," "team leaders," etc.)? Even though they are not officially "authorized" to take complaints, you may want to make sure they know what to do if they receive one.

Is Your Content Complete?

Interrelated with accurate content is the requirement that the training must cover everything. Certainly, one would have difficulty using training as an affirmative defense if the training did not cover all aspects of the policy. This is another seemingly obvious rule that frequently is not followed. So what types of issues are most likely forgotten?

Off-premises conduct is one culprit. One reason for the failure to include off-premises conduct is that many policies don't discuss it. Many employers aren't aware of the omission or don't feel they have an obligation to monitor conduct outside of work. Consider this example. Let's say that the training does not cover off-premises conduct. The employer then faces a situation in which part or all of the harassing conduct occurred outside the workplace, but with a work-related connection, for example, on a business trip or a sales conference. The training may be little help in showing "reasonable care" if it was silent on the issue. It is certainly okay, and in many instances a good practice, to address relevant issues that are not in the policy.

Another common problem we have encountered is a failure to train managers/supervisors about their affirmative duty to report harassment. This is most often caused by employees and managers and supervisors being trained together, rather than separately. Consider this scenario: Managers and supervisors are trained on harassment but are not trained on their obligation to report harassment up to a proper party to start an investigation. This obligation should be covered if managers and supervisors are named as agents for reporting in the policy. Later, employees are trained on the policy, which indicates that complaints of harassment can be made to managers and supervisors. An employee makes a complaint to a manager who, not knowing his obligations to report up, responds, "Boys will be boys. That's just the way Harry is. You'll get used to it." And, needless to say, the manager does nothing. How easy will it be to attack your "reasonable care" at trial? Very easy.

Speaking of an affirmative duty to report harassment, what about the trainer of the course? Do not forget that you, whoever you are (not personally, we mean within the organization), may also have an affirmative obligation to report harassment. We mention that because on numerous occasions a participant has come up to us after the class and mentioned that he or she is experiencing harassment. Even if the participant says, "I do not want to report this," you, like any manager or supervisor, may be considered to be legally on notice of harassment. The safest thing to do is to make sure that you pass that information on to the appropriate person(s) for investigation.

We also have observed many an employer fail to train on the process for reporting retaliation. This usually is a simple oversight by the trainer, generally caused by either discussing reporting procedures before retaliation or just mere forgetfulness on the trainer's part. It is simple to mention that the reporting procedures should be utilized for both harassment and retaliation complaints. It is also simple to forget.

And unquestionably, the flag-bearer of all complete content failures is to think of training simply as sexual harassment training. One more time, with feeling—*training must also be about harassment based on all protected characteristics.* Never forget race, color, religion, national origin, age, disability, or retaliation. Good. Enough said.

Outside of the multitude of altruistic reasons for conducting harassment training, the key reason for employers to conduct training is to help avoid legal liability. How ironic then that harassment training sessions have been known to directly lead to legal liability. The remaining portion of this chapter will focus on the inherent legal risks associated with harassment training.

 TIPS: ■ Audit your content against the EEOC Enforcement Guidance to assure your content is complete.

Are You Giving Legal Advice or Legal Conclusions?

Never give legal conclusions is a firm rule for training, whether or not you are a lawyer. Why? There is nothing worse than the following training scenario: Q: "If someone does X, Y, and Z, would that be harassment?" A: "Yes, of course it would." Problem: If X, Y, and Z are actually happening (and sometimes they are, that's why the question was asked) you have just made a legal conclusion. If the organization is sued for the X, Y, and Z that happened, you may be bound by your (damning) conclusion that this conduct violated the law. In fact, those conclusions are only for judges and juries to make. Trainers must focus on the policy, not on the law. Correct answer without falling into a pothole: "I can't say whether or not that conduct violates the law. Only a judge or jury can draw those legal conclusions. However, it does/may/can violate our policy and that is what we are focused on today."

Never give legal advice is the other part of this rule. If you are a lawyer, training is not the place to be dispensing legal advice. There will be no attorney–client privilege that attaches, particularly since you will want to be able to use the training as evidence of your affirmative defense.

 TIPS: ▪ We often tell participants that legal advice is a lot of facts and a lot of research (and a lot of money!). When we hear a lot of facts ("then she said to him, and then he said to her, and then they went out, and then . . . "), we invoke the no-legal-advice ground rule.

Are You Debating the Laws/Debating the Policy?

A training session should not discuss the merits of either discrimination laws or the policy. On a practical level, discussion of these controversial topics will inevitably lead to argument and debate, which will result in the loss of time and energy and in resistance to learning. On a more dangerous level, debating laws or the policy can result in a complete rejection of the policy by the participants. For debate and argument breed dissension. The trainer must decree that the policy is the law of the land. There is no debate. There is no discussion. There is no good or bad, right or wrong. Instead, the policy is part of the house rules of the employer. The central message must be: If you want to work here, you have to abide by these rules, regardless of your opinion of their merit or lack thereof.

Furthermore, discussing Title VII, or other laws, lends itself to using language that is not understandable, potentially creating a training that is not comprehensible and certainly not memorable. Did we mention boring?

 TIPS: ▪ This issue, as well as a number of issues reviewed here, is best handled by setting ground rules in the introductory phase of the training. (See Chapter 6 and discussion on ground rules.) We always tell participants that we are not there to debate the law or the policy, but that a better use of our time together is to understand exactly what it says and what is required. Then, if you need to later, you can return to that concept, and remind participants of that ground rule.

Are You Discussing Past, Current, or Pending Situations?

We have a firm rule that there will be no discussion of actual situations that have occurred in the workplace. Be careful with this rule, especially since participants will regularly want to discuss situations that have happened to them or someone they know in the workforce. In fact, not discussing situations in the workplace is somewhat counterintuitive, as they would probably provide the most relevant examples. But this rule is used to prevent legal problems associated with confidentiality, defamation, slander, and invasion of privacy issues. Discussing actual harassment situations with a roomful of employees can only lead to legal headaches. Not to mention the loss of credibility the employer will have when it ensures confidentiality to the greatest extent possible when conducting a harassment investigation.

In addition, do not use real-life situations that occurred outside the workplace as examples, unless you have a legitimate source to cite, like a reported case or a reputable newspaper. We also recommend against using company names, especially competitor names, even if it is just for a hypothetical. Hypothetical scenarios are sometimes misunderstood as factual, and one never knows who is in the room.

Are You Testing?

Don't do it on paper. A lot of training uses testing as a way of measuring a participant's mastery of the topic and their progress to completion. The problem with testing in harassment training boils down to whether you want to defend your decision to "pass" a participant who gets a low score. For example, if you give a test of any kind and you score it, what is the "cutoff" point for passing and failing? Is it 70 or 85? And if a participant gets an 85, and later is accused of harassing someone, do you think that the alleged harasser's "score" will be used against you? Absolutely. Let's peek in a courtroom already in session: "Did you do any remedial training? No? Why not? Were you satisfied that your workforce had mastered only 85 percent of the content? Didn't you worry about the other 15 percent? You tolerate 15 percent harassment?" So goes the cross-examination.

 TIPS: ▪ We use True/False cards for this reason (see Chapter 5). The trainer can see very clearly if anyone does not have the right answer when reviewing at the close of the training. Questions that participants still do not understand will be apparent to you and you can take the opportunity to make sure that no one leaves with any wrong answers or misunderstandings.

Do You Permit Discriminatory/Stereotyping Remarks?

At first blush, this seems like a rather obvious rule. Of course you wouldn't . . . let one slip in. But when leading a class on harassment, the chance of some stereotypical comments being made by participants is great and the chance of a trainer's example of discrimination being taken out of context is also risky. When your training is about overcoming discrimination, there are bound to be inappropriate remarks made, regardless of the intent.

For example, consider the case of *Stender v. Lucky Stores, Inc.,*[6] where "diversity/harassment" training for managers and supervisors factored in a reported settlement of more than $100 million. In that case the trainer had requested that each person at the training volunteer a stereotype about women. The resulting list included such statements as "women do not want to work late shifts because their husbands won't let them," "the crew won't work for a black woman," and "women seem to step down a lot after being promoted." These statements were offered as direct evidence in the subsequent "glass ceiling" class action suit. Whether the managers and supervisors believed in these stereotypes was unclear. But the court posited that many of these views reflected the views of at least some of the managers and supervisors. These questions begged for stereotyping. So, be careful what you ask for.

What a trainer must especially guard against, therefore, is discriminatory remarks or stereotypes made by managers or supervisors—the very group, after all, that has the power to act for the employer in making employment decisions. Remember what we said about attorney–client privilege? It may bear repeating. There is no privilege in training. That means, what is said in training can be used against the organization. This is particularly true for comments by managers and supervisors. Thus, the instant a participant begins to make inappropriate remarks the trainer must put a stop to it. Further, the trainer should never ask any participant about his/her personal feelings concerning discrimination or stereotypes, or about inappropriate behavior in the workplace. If it is necessary to discuss stereotypes or discriminatory behavior, the safest approach for trainers is to discuss society's stereotypes. Never make it personal—no matter how cathartic it may seem or how appropriate from a training perspective. Never, ever.

TIPS: ▪ If you are debriefing an exercise such as a role play, write out your questions ahead so that you do not thoughtlessly ask the audience questions that will elicit discriminatory responses.

Do You Use Offensive Language or Conduct?

The fact of the matter is that offensive language sometimes must be used and examples of harassment discussed to properly train about anti-harassment. Such language must not, however, be perceived as condoned by the employer. As a trainer, you should emphasize up front that any offensive language or offensive situations are being used only to help the participants better understand the policy. Then use a great deal of judgment before using sexual or racist language or describing extremely graphic situations.

We have heard of numerous instances in which a participant has later complained that they were offended by the training, or even filed a harassment complaint over it. There is little a trainer can do about such situations; it's simply an occupational hazard of the job. But by warning participants up front, not appearing flippant, not using outrageously offensive words such as nicknames for religions or races, and avoiding touching participants, even as demonstrations, a trainer will at least lessen the chance that a participant will be offended.

Do You Use Humor?

As we have mentioned before, the use of interactive training lends itself easily to humor. When done properly, participants will not only learn about the policy but enjoy themselves and, most important, retain what they learn. From our perspective, interactive training really requires humor and a "light touch." Yet some people feel that humor should never be used when dealing with serious subjects. We like to establish in the beginning that when we use humor, it does not mean that we do not take the topic seriously.

Are You Promising Confidentiality?

We have spent the last few pages warning about legal pitfalls to anti-harassment training. All of these pitfalls are based on the awful truth that there is no such thing as confidentiality. Like the misperception that employees often have regarding complete confidentiality during investigations, so too do people have misperceptions with respect to training sessions in general. Let's make this perfectly clear. Just because someone says something is confidential doesn't make it so.

We have heard stories of a trainer or employer being stunned because evidence is gathered from the training session even though the trainer and the participants "agreed" that everything said in the training stays in the training room, or is confidential. Remember, if someone is deposed and asked about the training session, the person cannot refuse to answer; he or she cannot say, "Pardon me, I took a pledge of confidentiality at the training and cannot speak." That won't work.

Anything said or done at the training is fair game to be discussed from the boardroom to the courtroom, even if the trainer is an attorney. There is no such thing as attorney–participant confidentiality. Most significantly, making a training confidential defeats the primary purpose of the training. An employer wants to use the training and what was taught at the training to establish the affirmative defense—to illustrate its reasonable efforts to prevent harassment or discrimination.

Are You Documenting?

In any legal issue, the primary purpose of documentation is to provide credible and material evidence in a legal dispute. And this is no different for harassment training. The ability to support training with documentation is critical to an employer's use of the affirmative defense. Without proper documentation, an employer's training efforts may be rendered useless in a court of law.

Generally, for our purposes, documents are needed to help prove that training was actually provided, to specify the content of the training, and to indicate which employees participated in the session. The goal is for documents to help show that the employer exercised reasonable care to prevent and correct promptly any harassing behavior—i.e., to provide evidence to establish the affirmative defense. Here are the types of documents that may be prepared when conducting harassment training.[7]

Attendance Sheets

As mentioned before, it is critical that every member of an organization be trained. There can be no exceptions. Attendance sheets are necessary to assist an employer in tracking who has taken (and not taken) training. The ultimate utility of the attendance sheet, however, is during litigation as evidence that the employer took reasonable care and trained the relevant parties, as well as the entire workforce.

In the never-ending quest to make life easier for your attorney(s), it is important that attendance sheets include as much information as possible pertaining to the logistics of the training. Thus, at minimum, attendance sheets should specify the title of the course, the date of the course, the location of the course, and the name(s) of the trainer(s). Also, it is prudent to have each participant both print and sign their name so that names can be easily ascertained at a later date. This will also minimize the need to hire hieroglyphics experts.

In large organizations, it may also be necessary to address the possibility of having a number of employees with the same name. Thus, to ensure accuracy, it is wise for an attendance form to require some type of identification, such as an employee identification number or a social security number. If not a social security number or employee identification number, any other identifying information is appropriate. But beware of asking for "risky" information, like an employee's date of birth.

Attach materials or a course outline to the attendance sheet before the sheet is filed away. This is especially wise if there have been several versions of the anti-harassment course, and to differentiate between manager/supervisor and employee courses.

In addition, the attendance sheets should be signed at the end of the training session by the participants. This way participants cannot get credit for taking the training if they leave the session early.

Finally, this tracking tool should be kept in the HR department and not in employee personnel files.

Participation Form

Closely related to attendance sheets, participation forms also are used to track who has taken training. Rather than one document signed by all participants, the participation form is given separately to each participant. Essentially, participation forms can be thought of as personal or private attendance sheets. They should be handed out at the end of the training and require each participant to print and sign his or her name as well as note the date of the training.

But unlike attendance sheets, a participation form needs to be kept in the employee's personnel file. And unlike attendance sheets, it is not necessary to specify the name of the trainer or the location of the class on the form because this information could easily be cross-referenced with the attendance sheets.

Policy Acknowledgment Form

There is no more critical form than the policy acknowledgment form. The ability of the employer to document that an employee admitted receiving and "reading" the anti-harassment policy will go a long way toward protecting the employer. While no one can be assured whether the policy is actually read, it can be assured that the policy is distributed to every employee. This is what is of the utmost importance, as illustrated by the *Faragher* case, where the key issue was the city's failure to disseminate its sexual harassment policy.

Thus, a good policy acknowledgment form locks in the fact that a participant has received the policy. It should also assert that the participant read the policy and understands it. This puts the onus on the participant to later argue that the policy was not read or understood. It is also helpful to include that the participant will abide by the terms and conditions of the policy, and face disciplinary actions, including termination, if the participant is found to have violated the policy.

Unlike attendance sheets and participation forms, a policy acknowledgment form should be handed out during the session, immediately following a discussion and review of the policy. This will hopefully not be the first time a participant signs such a form, as the employer should have employees sign such forms immediately after hire. Regardless of the time span between hiring and training, however, the participant should always sign the form.

Because the policy acknowledgment form is concerned primarily with the receipt of the policy, the form need not concern itself with the logistics of the training. Essentially, all that is needed is for the participant to print his/her name and sign a piece of paper that declares he or she has received, read, and understood the policy and will abide by it. The form should be placed in the participant's personnel file.

TIPS: ▪   The acknowledgment form should be separate and apart from the actual policy. Often we encounter a policy acknowledgment form that is on the same page as the policy itself. And when a participant is asked to turn in the form, the participant also tears away part of the policy. Sometimes the portion torn away includes the names and telephone numbers of the people to whom complaints should be made. It goes without saying that the policy always should be kept intact.

▪   Obviously, if you are handing out other policies or important materials during the training session, acknowledgment forms should be prepared and collected for each.

Evaluation Forms

Many trainers use optional evaluation forms for two primary reasons: to evaluate the trainer(s) and to evaluate the training. If the choice is made to use evaluation forms, there are a few items that should be considered. First and foremost, evaluation forms are discoverable. So one must weigh the benefit of using evaluation forms with the risk that participants may put things on the form that could prove extremely problematic. Obviously, comments like "I didn't learn anything from the training" or "the whole class, including myself, slept through the training" will make it more problematic to establish a defense based on training. Moreover, you will need to follow up with any notice issues such as, "I have been harassed here for years and no one has done anything." If the evaluations are signed, this will be easy to follow up. If they are anonymous, it is more difficult.

Another potential hazard is questions on the form that lend themselves to follow-up inquiries. For example, if an evaluation form asked, "Do you now understand that the organization has zero tolerance for harassment?" and the answer is "No," there may be a duty to retrain such participants. Most employers are not prepared to retrain scores of employees based on evaluation forms. This issue becomes even more discouraging when dealing with anonymous evaluation forms.

On the plus side, using evaluations will often provide the feedback needed to develop a quality training program. And, if the training is successful, the fact that evaluation forms provide evidence of such success will only boost an employer's legal defense. Further, positive evaluations can be used as an effective marketing tool for the training initiative.

Thus, we recommend using evaluation forms only if the trainer and the employer are completely comfortable with the training and expect positive feedback. Prior to this comfort level, trainers and employers should pursue feedback through informal methods that are not documented. We also recommend that if using evaluation forms, the forms be anonymous. Having participants sign the form will only chill the feedback. Finally, be careful in designing the form to ensure against placing the employer in the costly and dangerous position of retraining participants.

Documentation Dilemmas

When dealing with all forms, it is important to remember a few cardinal rules. Copies should be made of all signed forms. Copies should be kept in a separate location. There should be clear protocol as to the placement of all forms, both original and copies. Frequently, there are different versions or generations of the training and the policy. Therefore, documents must be updated and properly titled to warrant that the document signed was actually in reference to the policy read or the training taken.

One issue that arises with respect to documentation is a participant's refusal to sign the form. Handling a refusal can be tricky. While the goal is to have participants sign all forms, the trainer doesn't want to push too much and potentially lose control of the session. First you must determine if the participant refuses to sign because he or she does not understand something about the policy. If so, resolve that issue. Assuming plain old resistance, there are two quality tactics a trainer can take. One is to tell the participant to write on the form that "[name of participant] refuses to sign the form." The second option is for the trainer to tell the participant to simply print his/her name on the form rather than sign it. Both of these tactics will more often than not placate the resisting party, while still providing evidence that he or she was in the class and/or had the opportunity to read the policy. Of course, if someone refuses to put any mark on a form, the trainer should merely note on the form that the participant refused to sign. The trainer should also figure out who the participant is to later inform the employer. Also, the trainer should tell the participant that a failure to sign the policy acknowledgment form does not nullify his or her responsibilities under it.

If a trainer feels there is potential for an especially resistant class, the trainer should obtain a printout of the class and take roll call before the end of the training so that the trainer can document who was or was not in the class if any participant refuses to sign the attendance sheets or any other form.

TIPS:
- And as a friendly reminder, the actual curriculum or content of the training must always be documented. In litigation, the first issue that will arise with respect to training is the content of the training—what was taught. The better documented and organized the content of the training, the fewer headaches a trainer and an employer will face if litigation ensues.
- Finally, employers and trainers should be prepared to retain all documents pertaining to training for a substantial period of time. Your attorney will be less than impressed to hear that comprehensive training took place but there is no hard evidence/documents to support or prove it. How long should documents be kept? Well, there is no definitive answer. But with a recent Supreme Court decision potentially extending the time to sue for hostile environment claims,[8] we recommend retaining all documents pertaining to the training for a minimum of five years.

LESSONS LEARNED

To: What Could Go Wrong?
From: The Trainers
Re: Lessons Learned

Dear What Could Go Wrong?

Almost anything! We have given you an audit list of a baker's dozen questions to ask yourself in advance so you can avoid common pitfalls:

1. Have you trained from top to bottom?
2. Is your content accurate?
3. Are your reporting procedures accurate?
4. Is your content complete?
5. Are you giving legal advice or legal conclusions?
6. Are you debating the law/the policy?
7. Do you discuss past, current, or pending situations?
8. Are you testing?
9. Do you permit discriminatory/stereotyping remarks?
10. Do you use offensive language?
11. Do you use humor?
12. Are you promising confidentiality?
13. Are you documenting?

If you audit yourself against this list, you will improve your chances of having a successful session, from both a training and a legal perspective. And then, yes, indeed, life is good!

Signed,

The Trainers

Notes

1. An illustration of this is recent EEOC interrogatories that demand information about an employer's training program, including questions about—

- the scope of the training;
- the identities, qualification, and training of all individuals responsible for ensuring compliance with discrimination laws;
- the resources available and/or utilized for training; and
- the actual training material.

See Appendix E.

2. Besides, upper-level executives need to learn about their policies as much as, if not more than, the rank and file. *See, e.g.,* Mathis v. Phillips Chevrolet Inc., 269 F.3d 771 (7th Cir. 2001) (the general manager of a car dealership testified that he was not aware that it was illegal to discriminate based on age).

3. *See, e.g.,* Mallinson-Montague v. Pocrnick, 2000 WL 1346235 (10th Cir. 2000) (an alter ego instruction is appropriate against an employer based on allegations of sexual harassment committed by a manager or supervisor in those situations where the agent's high rank in the company makes him or her the employer's alter ego).

4. Cadena v. Pacesetter Corp., 224 F.3d 1203 (10th Cir. 2000).

5. *See* Gentry v. Export Packaging Co., 238 F.3d 842, 847–48 (7th Cir. 2001) (affirming jury verdict for plaintiff; policy was arguably deficient where it failed to designate who in human resources department was authorized to receive reports of harassment, and where an undertrained low-level HR representative failed to recognize the complaint as such because the plaintiff did not use the term "sexual harassment"); Smith v. First Union Nat'l Bank, 202 F.3d 234, 244–46 (4th Cir. 2000) (reversing summary judgment against sexual harassment claim; employer cannot establish the affirmative defense as a matter of law where the employer's investigator had never done one before, never asked alleged harasser whether he made the comments alleged, and simply counseled him to improve his management style and "smile more").

6. 803 F. Supp. 259 (N.D. Cal. 1992).

7. *See* Appendix K–N for sample documents.

8. Amtrak v. Morgan, 122 S.Ct. 2061 (2002) (court held that in hostile work environment claims, a plaintiff may sue over otherwise time-barred incidents, as long as one of the alleged incidents occurred within the charge-filing period, subject to equitable defenses).

APPENDIX A

Beth Ann Faragher, Petitioner v. City of Boca Raton

U.S. SUPREME COURT

Syllabus

v.

CERTIORARI TO THE UNITED STATES COURT OF APPEALS FOR THE ELEVENTH CIRCUIT

No. 97-282.
Argued March 25, 1998–
Decided June 26, 1998

After resigning as a lifeguard with respondent City of Boca Raton (City), petitioner Beth Ann Faragher brought an action against the City and her immediate supervisors, Bill Terry and David Silverman, for nominal damages and other relief, alleging, among other things, that the supervisors had created a "sexually hostile atmosphere" at work by repeatedly subjecting Faragher and other female lifeguards to "uninvited and offensive touching," by making lewd remarks, and by speaking of women in offensive terms, and that this conduct constituted discrimination in the "terms, conditions, and privileges" of her employment in violation of Title VII of the Civil Rights Act of 1964, 42 U.S.C. § 2000e-2(a)(1). Following a bench trial, the District Court concluded that the supervisors' conduct was discriminatory harassment sufficiently serious to alter the conditions of Faragher's employment and constitute an abusive working environment. The District Court then held that the City could be held liable for the harassment of its supervisory employees because the harassment was pervasive enough to support an inference that the City had "knowledge, or constructive knowledge" of it; under traditional agency principles Terry and Silverman were acting as the

City's agents when they committed the harassing acts; and a third supervisor had knowledge of the harassment and failed to report it to City officials. The Eleventh Circuit, sitting en banc, reversed. Relying on Meritor Savings Bank, FSB v. Vinson, 477 U.S. 57, and on the Restatement (Second) of Agency § 219 (1957) (Restatement), the Court of Appeals held that Terry and Silverman were not acting within the scope of their employment when they engaged in the harassing conduct, that their agency relationship with the City did not facilitate the harassment, that constructive knowledge of it could not be imputed to the City because of its pervasiveness or the supervisor's knowledge, and that the City could not be held liable for negligence in failing to prevent it.

Held: An employer is vicariously liable for actionable discrimination caused by a supervisor, but subject to an affirmative defense looking to the reasonableness of the employer's conduct as well as that of the plaintiff victim. Pp. 7–32.

(a) While the Court has delineated the substantive contours of the hostile environment Title VII forbids, see, e.g., Harris v. Forklift Systems, Inc., 510 U.S. 17, 21–22, its cases have established few definitive rules for determining when an employer will be liable for a discriminatory environment that is otherwise actionably abusive. The Court's only discussion to date of the standards of employer liability came in Meritor, supra, where the Court held that traditional agency principles were relevant for determining employer liability. Although the Court cited the Restatement §§ 219–237 with general approval, the Court cautioned that common-law agency principles might not be transferable in all their particulars. Pp. 7–14.

(b) Restatement § 219(1) provides that "a master is subject to liability for the torts of his servants committed while acting in the scope of their employment." Although Title VII cases in the Court of Appeals have typically held, or assumed, that supervisory sexual harassment falls outside the scope of employment because it is motivated solely by individual desires and serves no purpose of the employer, these cases appear to be in tension with others defining the scope of the employment broadly to hold employers vicariously liable for employees' intentional torts, including sexual assaults, that were not done to serve the employer, but were deemed to be characteristic of its activities or a foreseeable consequence of its business. This tension is the result of differing judgments about the desirability of holding an employer liable for his subordinates' wayward behavior. The proper analysis here, then, calls not for a mechanical application of indefinite and malleable factors set forth in the Restatement, but rather an enquiry into whether it is proper to conclude that sexual harassment is one of the normal risks of doing business the employer should bear. An employer can reasonably anticipate the possibility of sexual harassment occurring in the workplace, and this might justify the assignment of the costs of this behavior to the employer rather than to the victim. Two things counsel in favor of the contrary conclusion, however. First, there is no reason to suppose that Congress wished courts to ignore the traditional distinction between acts falling within the scope of employment and acts amounting to what the older law called frolics or detours from the course of employment. Second, the lower courts, by uniformly judging employer liability for co-worker harassment under a negligence standard, have implicitly treated such harassment outside the scope of employment. It is unlikely that such treatment would escape efforts to render them obsolete if the Court held that harassing supervisors necessarily act within the scope of their employment. The rationale for doing so would apply when the behavior was that of co-employees, because the employer generally benefits from the work of common employees as from the work of supervisors. The answer to this argument might be that the scope of supervisory employment may be treated separately because supervisors have special authority enhancing their capacity to harass and the employer can guard against their misbehavior more easily. This answer, however, implicates an entirely separate category of agency law, considered in the next section. Given the virtue of categorical clarity, it is better to reject reliance on misuse of supervisory authority (without more) as irrelevant to the scope-of-employment analysis. Pp. 14–23.

(c) The Court of Appeals erred in rejecting a theory of vicarious liability based on § 219(2)(d) of the Restatement, which provides that an employer "is not subject to liability for the torts of his servants acting outside the scope of their employment unless . . . the servant purported to act or speak on behalf of the principal and there was reliance on apparent authority, or he was aided in accomplishing the tort by the existence of the agency relation." It makes sense to hold an employer vicariously liable under Title VII for some tortious conduct of a supervisor made possible by use of his supervisory authority, and the aided-by-agency-relation principle of § 219(2)(d) provides an appropriate starting point for determining liability for the kind of harassment presented here. In a sense a supervisor is always assisted in his misconduct by the supervisory relationship; however, the imposition of liability based on the misuse of supervisory authority must be squared with Meritor's holding that an employer is not "automatically" liable for harassment by a supervisor who creates the requisite degree of discrimination. There are two basic alternatives to counter the risk of automatic liability. The first is to require proof of some affirmative invocation of that authority by the harassing supervisor; the second is to recognize an affirmative defense to liability in some circumstances, even when a supervisor has created the actionable environment. The problem with the first alternative is that there is not a clear line between the affirmative and merely implicit uses of supervisory power; such a rule would often lead to close judgment calls and results that appear disparate if not contradictory, and the temptation to litigate would be hard to resist. The second alternative would avoid this particular temptation to litigate and implement Title VII sensibly by giving employers an incentive to prevent and eliminate harassment and by requiring employees to take advantage of the preventive or remedial apparatus of their employers. Thus, the Court adopts the following holding in this case and in Burlington

Industries, Inc. v. Ellerth, p. ___, also decided today. An employer is subject to vicarious liability to a victimized employee for an actionable hostile environment created by a supervisor with immediate (or successively higher) authority over the employee. When no tangible employment action is taken, a defending employer may raise an affirmative defense to liability or damages, subject to proof by a preponderance of the evidence. See Fed. Rule Civ. Proc. 8(c). The defense comprises two necessary elements: (a) that the employer exercised reasonable care to prevent and correct promptly any sexually harassing behavior, and (b) that the plaintiff employee unreasonably failed to take advantage of any preventive or corrective opportunities provided by the employer or to avoid harm otherwise. While proof that an employer had promulgated an antiharassment policy with complaint procedure is not necessary in every instance as a matter of law, the need for a stated policy suitable to the employment circumstances may appropriately be addressed in any case when litigating the first element of the defense. And while proof that an employee failed to fulfill the corresponding obligation of reasonable care to avoid harm is not limited to showing an unreasonable failure to use any complaint procedure provided by the employer, a demonstration of such failure will normally suffice to satisfy the employer's burden under the second element of the defense. No affirmative defense is available, however, when the supervisor's harassment culminates in a tangible employment action, such as discharge, demotion, or undesirable reassignment. Pp. 23–30.

(**d**) Under this standard, the Eleventh Circuit's judgment must be reversed. The District Court found that the degree of hostility in the work environment rose to the actionable level and was attributable to Silverman and Terry, and it is clear that these supervisors were granted virtually unchecked authority over their subordinates and that Faragher and her colleagues were completely isolated from the City's higher management. While the City would have an opportunity to raise an affirmative defense if there were any serious prospect of its presenting one, it appears from the record that any such avenue is closed. The District Court found that the City had entirely failed to disseminate its sexual harassment policy among the beach employees and that its officials made no attempt to keep track of the conduct of supervisors, and the record makes clear that the City's policy did not include any harassing supervisors assurance that could be bypassed in registering complaints. Under such circumstances, the Court holds as a matter of law that the City could not be found to have exercised reasonable care to prevent the supervisors' harassing conduct. Although the record discloses two possible grounds upon which the City might seek to excuse its failure to distribute its policy and to establish a complaint mechanism, both are contradicted by the record. The City points to nothing that might justify a conclusion by the District Court on remand that the City had exercised reasonable care. Nor is there any reason to remand for consideration of Faragher's efforts to mitigate her own damages, since the award to her was solely nominal. Pp. 30–32.

(e) There is no occasion to consider whether the supervisors' knowledge of the harassment could be imputed to the City. Liability on that theory could not be determined without further factfinding on remand, whereas the reversal necessary on the supervisory harassment theory renders any remand for consideration of imputed knowledge (or of negligence as an alternative to a theory of vicarious liability) entirely unjustifiable. P. 32.

111 F. 3d 1530, reversed and remanded.

SOUTER, J., delivered the opinion of the Court, in which REHNQUIST, C. J., and STEVENS, O'CONNOR, KENNEDY, GINSBURG, and BREYER, J J., joined. THOMAS, J., filed a dissenting opinion, in which SCALIA, J., joined.

U.S. Supreme Court

No. 97-282

BETH ANN FARAGHER, PETITIONER v. CITY OF BOCA RATON

ON WRIT OF CERTIORARI TO THE UNITED STATES
COURT OF APPEALS FOR THE ELEVENTH CIRCUIT

[June 26, 1998]

JUSTICE SOUTER delivered the opinion of the Court.

This case calls for identification of the circumstances under which an employer may be held liable under Title VII of the Civil Rights Act of 1964, 78 Stat. 253, as amended, 42 U.S.C. § 2000e et seq., for the acts of a supervisory employee whose sexual harassment of subordinates has created a hostile work environment amounting to employment discrimination. We hold that an employer is vicariously liable for actionable discrimination caused by a supervisor, but subject to an affirmative defense looking to the reasonableness of the employer's conduct as well as that of a plaintiff victim.

I

Between 1985 and 1990, while attending college, petitioner Beth Ann Faragher worked part time and during the summers as an ocean lifeguard for the Marine Safety Section of the Parks and Recreation Department of respondent, the City of Boca Raton, Florida (City). During this period, Faragher's immediate supervisors were Bill Terry, David Silverman, and Robert Gordon. In June 1990, Faragher resigned.

In 1992, Faragher brought an action against Terry, Silverman, and the City, asserting claims under Title VII, 42 U.S.C. § 1983 and Florida law. So far as it concerns the Title VII claim, the complaint alleged that Terry and Silverman created a "sexually hostile atmosphere" at the beach by repeatedly subjecting Faragher and other female lifeguards to "uninvited and offensive touching," by making lewd remarks, and by speaking of women in offensive terms. The complaint contained specific allegations that Terry once said that he would never promote a woman to the rank of lieutenant, and that Silverman had said to Faragher, "Date me or clean the toilets for a year." Asserting that Terry and Silverman were agents of the City, and that their conduct amounted to discrimination in the "terms, conditions, and privileges" of her employment, 42 U.S.C. § 2000e-2(a)(1), Faragher sought a judgment against the City for nominal damages, costs, and attorney's fees.

Following a bench trial, the United States District Court for the Southern District of Florida found that throughout Faragher's employment with the City, Terry served as Chief of the Marine Safety Division, with authority to hire new lifeguards (subject to the approval of higher management), to supervise all aspects of the lifeguards' work assignments, to engage in counseling, to deliver oral reprimands, and to make a record of any such discipline. 864 F. Supp. 1552, 1563–1564 (1994). Silverman was a Marine Safety lieutenant from 1985 until June 1989, when he became a captain. Id., at 1555. Gordon began the employment period as a lieutenant and at some point was promoted to the position of training captain. In these positions, Silverman and Gordon were responsible for making the lifeguards' daily assignments, and for supervising their work and fitness training. Id., at 1564.

The lifeguards and supervisors were stationed at the city beach and worked out of the Marine Safety Headquarters, a small one-story building containing an office, a meeting room, and a single, unisex locker room with a shower. Id., at 1556. Their work routine was structured in a "paramilitary configuration," id., at 1564, with a clear chain of command. Lifeguards reported to lieutenants and captains, who reported to Terry. He was supervised by the Recreation Superintendent, who in turn reported to a Director of Parks and Recreation, answerable to the City Manager. Id., at 1555. The lifeguards had no significant contact with higher city officials like the Recreation Superintendent. Id., at 1564.

In February 1986, the City adopted a sexual harassment policy, which it stated in a memorandum from the City Manager addressed to all employees. Id., at 1560. In May 1990, the City revised the policy and reissued a statement of it. Ibid. Although the City may actually have circulated the memos and statements to some employees, it completely failed to disseminate its policy among employees of the Marine Safety Section, with the result that Terry, Silverman, Gordon, and many lifeguards were unaware of it. Ibid.

From time to time over the course of Faragher's tenure at the Marine Safety Section, between 4 and 6 of the 40 to 50 lifeguards were women. Id., at 1556. During that 5-year period, Terry repeatedly touched the bodies of female employees without invitation, ibid., would put his arm around Faragher, with his hand on her buttocks, id., at 1557, and once made contact with another female lifeguard in a motion of sexual simulation, id., at 1556. He made crudely demeaning references to women generally, id., at 1557, and once commented disparagingly on Faragher's shape, ibid. During a job interview with a woman he hired as a lifeguard, Terry said that the female lifeguards had sex with their male counterparts and asked whether she would do the same. Ibid. Silverman behaved in similar ways. He once tackled Faragher and remarked that, but for a physical characteristic he found unattractive, he would readily have had sexual relations with her. Ibid. Another time, he pantomimed an act of oral sex. Ibid. Within earshot of the female lifeguards, Silverman made frequent, vulgar references to women and sexual matters, commented on the bodies of female lifeguards and beachgoers, and at least twice told female lifeguards that he would like to engage in sex with them. Id., at 1557–1558.

Faragher did not complain to higher management about Terry or Silverman. Although she spoke of their behavior to Gordon, she did not regard these discussions as formal complaints to a supervisor but as conversations with a person she held in high esteem. Id., at 1559. Other female lifeguards had similarly informal talks with Gordon, but because Gordon did not feel that it was his place to do so, he did not report these complaints to Terry, his own supervisor, or to any other city official. Id., at 1559–1560. Gordon responded to the complaints of one lifeguard by saying that "the City just [doesn't] care." Id., at 1561.

In April 1990, however, two months before Faragher's resignation, Nancy Ewanchew, a former lifeguard, wrote to Richard Bender, the City's Personnel Director, complaining that Terry and Silverman had harassed her and other female lifeguards. Id., at 1559. Following investigation of this complaint, the City found that Terry and Silverman had behaved improperly, reprimanded them, and required them to choose between a suspension without pay or the forfeiture of annual leave. Ibid.

On the basis of these findings, the District Court concluded that the conduct of Terry and Silverman was discriminatory harassment sufficiently serious to alter the conditions of Faragher's employment and constitute an abusive working environment. Id., at 1562–1563. The District Court then ruled that there were three justifications for holding the City liable for the harassment of its supervisory employees. First, the court noted that the harassment was pervasive enough to support an inference that the City had "knowledge, or constructive knowledge" of it. Id., at 1563. Next, it ruled that the City was liable under traditional agency principles because Terry and Silverman were acting as its agents when they committed the harassing acts. Id., at 1563–1564. Finally, the court observed that Gordon's knowledge of the harassment, combined with his inaction, "provides a further basis for imputing liability on [sic] the City." Id., at 1564. The District Court then awarded Faragher one dollar in nominal damages on her Title VII claim. Id., at 1564–1565.

A panel of the Court of Appeals for the Eleventh Circuit reversed the judgment against the City. 76 F. 3d 1155 (1996). Although the panel had "no trouble concluding that Terry's and Silverman's conduct . . . was severe and pervasive enough to create an objectively abusive work environment," id., at 1162, it overturned the District Court's conclusion that the City was liable. The panel ruled that Terry and Silverman were not acting within the scope of their employment when they engaged in the harassment, id., at 1166, that they were not aided in their actions by the agency relationship, id., at 1166, n. 14, and that the City had no constructive knowledge of the harassment by virtue of its pervasiveness or Gordon's actual knowledge, id., at 1167, and n. 16.

In a 7-to-5 decision, the full Court of Appeals, sitting en banc, adopted the panel's conclusion. 111 F. 3d 1530 (1997). Relying on our decision in Meritor Savings Bank, FSB v. Vinson, 477 U.S. 57 (1986), and on the Restatement (Second) of Agency § 219 (1957) (hereafter Restatement), the court held that "an employer may be indirectly liable for hostile environment sexual harassment by a superior: (1) if the harassment occurs within the scope of the superior's employment; (2) if the employer assigns performance of a nondelegable duty to a supervisor and an employee is injured because of the supervisor's failure to carry out that duty; or (3) if there is an agency relationship which aids the supervisor's ability or opportunity to harass his subordinate." Id., at 1534–1535.

Applying these principles, the court rejected Faragher's Title VII claim against the City. First, invoking standard agency language to classify the harassment by each supervisor as a "frolic" unrelated to his authorized tasks, the court found that in harassing Faragher, Terry and Silverman were acting outside of the scope of their employment and solely to further their own personal ends. Id., at 1536–1537. Next, the court determined that the supervisors' agency relationship with the City did not assist them in perpetrating their harassment. Id., at 1537. Though noting that "a supervisor is always aided in accomplishing hostile environment sexual harassment by the existence of the agency relationship with his employer because his responsibilities include close proximity to and regular contact with the victim," the court held that traditional agency law does not employ so broad a concept of aid as a predicate of employer liability, but requires something more than a mere combination of agency relationship and improper conduct by the agent. Ibid. Because neither Terry nor Silverman threatened to fire or demote

Faragher, the court concluded that their agency relationship did not facilitate their harassment. Ibid.

The en banc court also affirmed the panel's ruling that the City lacked constructive knowledge of the supervisors' harassment. The court read the District Court's opinion to rest on an erroneous legal conclusion that any harassment pervasive enough to create a hostile environment must a fortiori also suffice to charge the employer with constructive knowledge. Id., at 1538. Rejecting this approach, the court reviewed the record and found no adequate factual basis to conclude that the harassment was so pervasive that the City should have known of it, relying on the facts that the harassment occurred intermittently, over a long period of time, and at a remote location. Ibid. In footnotes, the court also rejected the arguments that the City should be deemed to have known of the harassment through Gordon, id., at 1538, n. 9, or charged with constructive knowledge because of its failure to disseminate its sexual harassment policy among the lifeguards, id., at 1539, n. 11.

Since our decision in Meritor, Courts of Appeals have struggled to derive manageable standards to govern employer liability for hostile environment harassment perpetrated by supervisory employees. While following our admonition to find guidance in the common law of agency, as embodied in the Restatement, the Courts of Appeals have adopted different approaches. Compare, e.g., Harrison v. Eddy Potash, Inc., 112 F. 3d 1437 (CA10 1997), cert. pending, No. 97-232; 111 F. 3d 1530 (CA11 1997) (case below); Gary v. Long, 59 F. 3d 1391 (CADC), cert. denied, 516 U.S. 1011 (1995); and Karibian v. Columbia University, 14 F. 3d 773 (CA2), cert. denied, 512 U.S. 1213 (1994). We granted certiorari to address the divergence, 522 U.S. ___ (1997), and now reverse the judgment of the Eleventh Circuit and remand for entry of judgment in Faragher's favor.

II

A

Under Title VII of the Civil Rights Act of 1964, "[i]t shall be an unlawful employment practice for an employer . . . to fail or refuse to hire or to discharge any individual, or otherwise to discriminate against any individual with respect to his compensation, terms, conditions, or privileges of employment, because of such individual's race, color, religion, sex, or national origin." 42 U.S.C. § 2000e2(a)(1). We have repeatedly made clear that although the statute mentions specific employment decisions with immediate consequences, the scope of the prohibition " 'is not limited to "economic" or "tangible" discrimination,' " Harris v. Forklift Systems, Inc., 510 U.S. 17, 21 (1993) (quoting Meritor Savings Bank, FSB v. Vinson, 477 U.S., at 64), and that it covers more than " 'terms' and 'conditions' in the narrow contractual sense." Oncale v. Sundowner Offshore Services, Inc., 523 U.S. ___ ,

___ (1998) (slip op., at 2–3). Thus, in Meritor we held that sexual harassment so "severe or pervasive" as to " 'alter the conditions of [the victim's] employment and create an abusive working environment' " violates Title VII. 477 U.S., at 67 (quoting Henson v. Dundee, 682 F. 2d 897, 904 (CA11 1982)).

In thus holding that environmental claims are covered by the statute, we drew upon earlier cases recognizing liability for discriminatory harassment based on race and national origin, see, e.g., Rogers v. EEOC, 454 F. 2d 234 (CA5 1971), cert. denied, 406 U.S. 957 (1972); Firefighters Institute for Racial Equality v. St. Louis, 549 F. 2d 506 (CA8), cert. denied sub nom. Banta v. United States, 434 U.S. 819 (1977), just as we have also followed the lead of such cases in attempting to define the severity of the offensive conditions necessary to constitute actionable sex discrimination under the statute. See, e.g., Rogers, supra, at 238 ("[M]ere utterance of an ethnic or racial epithet which engenders offensive feelings in an employee" would not sufficiently alter terms and conditions of employment to violate Title VII).[1]

See also Daniels v. Essex Group, Inc., 937 F. 2d 1264, 1271–1272 (CA7 1991); Davis v. Monsanto Chemical Co., 858 F. 2d 345, 349 (CA6 1988), cert. denied, 490 U.S. 1110 (1989); Snell v. Suffolk County, 782 F. 2d 1094, 1103 (CA2 1986); 1 B. Lindemann & P. Grossman, Employment Discrimination Law 349, and nn. 36–37 (3d ed. 1996) (hereinafter Lindemann & Grossman) (citing cases instructing that "[d]iscourtesy or rudeness should not be confused with racial harassment" and that "a lack of racial sensitivity does not, alone, amount to actionable harassment").

So, in Harris, we explained that in order to be actionable under the statute, a sexually objectionable environment must be both objectively and subjectively offensive, one that a reasonable person would find hostile or abusive, and one that the victim in fact did perceive to be so. 510 U.S., at 21–22. We directed courts to determine whether an environment is sufficiently hostile or abusive by "looking at all the circumstances," including the "frequency of the discriminatory conduct; its severity; whether it is physically threatening or humiliating, or a mere offensive utterance; and whether it unreasonably interferes with an employee's work performance." Id., at 23. Most recently, we explained that Title VII does not prohibit "genuine but innocuous differences in the ways men and women routinely interact with members of the same sex and of the opposite sex." Oncale, 523 U.S., at ___ (slip op., at 6). A recurring point in these opinions is that "simple teasing," id., at ___ (slip op., at 7), offhand comments, and isolated incidents (unless extremely serious) will not amount to discriminatory changes in the "terms and conditions of employment."

These standards for judging hostility are sufficiently demanding to ensure that Title VII does not become a "general civility code." Id., at ___ (slip op., at 6). Properly applied, they will filter out complaints attacking "the ordinary tribulations of the workplace, such as the sporadic use of abusive language, gender-related jokes, and occasional teasing." B. Lindemann & D. Kadue, Sexual Harassment in Employment Law 175 (1992) (hereinafter Lindemann & Kadue) (footnotes omitted). We have made it clear that conduct must be extreme to amount to a change in the terms and conditions of employment, and the Courts of Appeals have heeded this view. See, e.g., Carrero v. New York City Housing Auth., 890 F. 2d 569, 577578 (CA2 1989); Moylan v. Maries County, 792 F. 2d 746, 749–750 (CA8 1986); See also 1 Lindemann & Grossman 805–807, n. 290 (collecting cases granting summary judgment for employers because the alleged harassment was not actionably severe or pervasive).

While indicating the substantive contours of the hostile environments forbidden by Title VII, our cases have established few definite rules for determining when an employer will be liable for a discriminatory environment that is otherwise actionably abusive. Given the circumstances of many of the litigated cases, including some that have come to us, it is not surprising that in many of them, the issue has been joined over the sufficiency of the abusive conditions, not the standards for determining an employer's liability for them. There have, for example, been myriad cases in which District Courts and Courts of Appeals have held employers liable on account of actual knowledge by the employer, or high-echelon officials of an employer organization, of sufficiently harassing action by subordinates, which the employer or its informed officers have done nothing to stop. See, e.g., Katz v. Dole, 709 F. 2d 251, 256 (CA4 1983) (upholding employer liability because the "employer's supervisory personnel manifested unmistakable acquiescence in or approval of the harassment"); EEOC v. Hacienda Hotel, 881 F. 2d 1504, 1516 (CA9 1989) (employer liable where hotel manager did not respond to complaints about supervisors' harassment); Hall v. Gus Constr. Co., 842 F. 2d 1010, 1016 (CA8 1988) (holding employer liable for harassment by co-workers because supervisor knew of the harassment but did nothing). In such instances, the combined knowledge and inaction may be seen as demonstrable negligence, or as the employer's adoption of the offending conduct and its results, quite as if they had been authorized affirmatively as the employer's policy. Cf. Oncale, supra, at ___ (slip op., at 2) (victim reported his grounds for fearing rape to company's safety supervisor, who turned him away with no action on complaint).

Nor was it exceptional that standards for binding the employer were not in issue in Harris, supra. In that case of discrimination by hostile environment, the individual charged with creating the abusive atmosphere was the president of the corporate employer, 510 U.S., at 19, who was indisputably within that class of an employer organization's officials who may be treated as the organization's proxy. Burns v. McGregor Electronic Industries, Inc., 955 F. 2d 559, 564 (CA8 1992) (employer-company liable where harassment was perpetrated by its owner); see Torres v. Pisano, 116 F. 3d 625,

634–635, and n. 11 (CA2) (noting that a supervisor may hold a sufficiently high position "in the management hierarchy of the company for his actions to be imputed automatically to the employer"), cert. denied, 522 U.S. ___ (1997); cf. Katz, supra, at 255 ("Except in situations where a proprietor, partner or corporate officer participates personally in the harassing behavior," an employee must "demonstrat[e] the propriety of holding the employer liable").

Finally, there is nothing remarkable in the fact that claims against employers for discriminatory employment actions with tangible results, like hiring, firing, promotion, compensation, and work assignment, have resulted in employer liability once the discrimination was shown. See Meritor, 477 U.S., at 70–71 (noting that "courts have consistently held employers liable for the discriminatory discharges of employees by supervisory personnel, whether or not the employer knew, should have known, or approved of the supervisor's actions"); id., at 75 (Marshall, J., concurring in judgment) ("[W]hen a supervisor discriminatorily fires or refuses to promote a black employee, that act is, without more, considered the act of the employer"); see also Anderson v. Methodist Evangelical Hospital, Inc., 464 F. 2d 723, 725 (CA6 1972) (imposing liability on employer for racially motivated discharge by low-level supervisor, although the "record clearly shows that [its] record in race relations . . . is exemplary").

A variety of reasons have been invoked for this apparently unanimous rule. Some courts explain, in a variation of the "proxy" theory discussed above, that when a supervisor makes such decisions, he "merges" with the employer, and his act becomes that of the employer. See, e.g., Kotcher v. Rosa and Sullivan Appliance Ctr., Inc., 957 F. 2d 59, 62 (CA2 1992) ("The supervisor is deemed to act on behalf of the employer when making decisions that affect the economic status of the employee. From the perspective of the employee, the supervisor and the employer merge into a single entity"); Steele v. Offshore Shipbuilding, Inc., 867 F. 2d 1311, 1316 (CA11 1989) ("When a supervisor requires sexual favors as a quid pro quo for job benefits, the supervisor, by definition, acts as the company"); see also Lindemann & Grossman 776 (noting that courts hold employers "automatically liable" in quid pro quo cases because the "supervisor's actions, in conferring or withholding employment benefits, are deemed as a matter of law to be those of the employer"). Other courts have suggested that vicarious liability is proper because the supervisor acts within the scope of his authority when he makes discriminatory decisions in hiring, firing, promotion, and the like. See, e.g., Shager v. Upjohn Co., 913 F. 2d 398, 405 (CA7 1990) ("[A] supervisory employee who fires a subordinate is doing the kind of thing that he is authorized to do, and the wrongful intent with which he does it does not carry his behavior so far beyond the orbit of his responsibilities as to excuse the employer") (citing Restatement § 228). Others have suggested that vicarious liability is appropriate because the supervisor who discriminates in this manner is aided by the agency relation. See, e.g., Nichols v. Frank, 42 F. 3d 503, 514 (CA9 1994). Finally, still other courts have endorsed both of the latter two theories. See, e.g., Harrison, 112 F. 3d, at 1443; Henson, 682, F. 2d, at 910.

The soundness of the results in these cases (and their continuing vitality), in light of basic agency principles, was confirmed by this Court's only discussion to date of standards of employer liability, in Meritor, supra, which involved a claim of discrimination by a supervisor's sexual harassment of a subordinate over an extended period. In affirming the Court of Appeals's holding that a hostile atmosphere resulting from sex discrimination is actionable under Title VII, we also anticipated proceedings on remand by holding agency principles relevant in assigning employer liability and by rejecting three per se rules of liability or immunity. 477 U.S., at 70–72. We observed that the very definition of employer in Title VII, as including an "agent," id., at 72, expressed Congress's intent that courts look to traditional principles of the law of agency in devising standards of employer liability in those instances where liability for the actions of a supervisory employee was not otherwise obvious, ibid., and although we cautioned that "common-law principles may not be transferable in all their particulars to Title VII," we cited the Restatement §§ 219–237, with general approval. Ibid.

We then proceeded to reject two limitations on employer liability, while establishing the rule that some limitation was intended. We held that neither the existence of a company grievance procedure nor the absence of actual notice of the harassment on the part of upper management would be dispositive of such a claim; while either might be relevant to the liability, neither would result automatically in employer immunity. Ibid. Conversely, we held that Title VII placed some limit on employer responsibility for the creation of a discriminatory environment by a supervisor, and we held that Title VII does not make employers "always automatically liable for sexual harassment by their supervisors," ibid., contrary to the view of the Court of Appeals, which had held that "an employer is strictly liable for a hostile environment created by a supervisor's sexual advances, even though the employer neither knew nor reasonably could have known of the alleged misconduct," id., at 69–70. Meritor' s statement of the law is the foundation on which we build today. Neither party before us has urged us to depart from our customary adherence to stare decisis in statutory interpretation, Patterson v. McLean Credit Union, 491 U.S. 164, 172–173 (1989) (stare decisis has "special force" in statutory interpretation). And the force of precedent here is enhanced by Congress's amendment to the liability provisions of Title VII since the Meritor decision, without providing any modification of our holding. Civil Rights Act of 1991, § 102, 105 Stat. 1072, 42 U.S.C. § 1981a; see Keene Corp. v. United States, 508 U.S. 200, 212 (1993) (applying the "presumption that Congress was aware of [prior] judicial interpretations and, in effect, adopted them"). See also infra, at 26, n. 4.

B

The Court of Appeals identified, and rejected, three possible grounds drawn from agency law for holding the City vicariously liable for the hostile environment created by the supervisors. It considered whether the two supervisors were acting within the scope of their employment when they engaged in the harassing conduct. The court then enquired whether they were significantly aided by the agency relationship in committing the harassment, and also considered the possibility of imputing Gordon's knowledge of the harassment to the City. Finally, the Court of Appeals ruled out liability for negligence in failing to prevent the harassment. Faragher relies principally on the latter three theories of liability.

1

A "master is subject to liability for the torts of his servants committed while acting in the scope of their employment." Restatement § 219(1). This doctrine has traditionally defined the "scope of employment" as including conduct "of the kind [a servant] is employed to perform," occurring "substantially within the authorized time and space limits," and "actuated, at least in part, by a purpose to serve the master," but as excluding an intentional use of force "unexpectable by the master." Id., § 228(1).

Courts of Appeals have typically held, or assumed, that conduct similar to the subject of this complaint falls outside the scope of employment. See, e.g., Harrison, 112 F. 3d, at 1444 (sexual harassment " 'simply is not within the job description of any supervisor or any other worker in any reputable business' "); 111 F. 3d, at 1535–1536 (case below); Andrade v. Mayfair Management, Inc., 88 F. 3d 258, 261 (CA4 1996) ("[I]llegal sexual harassment is . . . beyond the scope of supervisors' employment"); Gary, 59 F. 3d, at 1397 (harassing supervisor acts outside the scope of his employment in creating hostile environment); Nichols v. Frank, 42 F. 3d 503, 508 (CA9 1994) ("The proper analysis for employer liability in hostile environment cases is . . . not whether an employee was acting within his 'scope of employment' "); Bouton v. BMW of North Am., Inc., 29 F. 3d 103, 107 (CA3 1994) (sexual harassment is outside scope of employment); see also Ellerth v. Burlington Industries, Inc., decided with Jansen v. Packaging Corp. of America, 123 F. 3d 490, 561 (CA7 1997) (en banc) (Manion, J., concurring and dissenting) (supervisor's harassment would fall within scope of employment only in "the rare case indeed"), cert. granted, No. 97-569; Lindemann & Grossman 812 ("Hostile environment sexual harassment normally does not trigger respondeat superior liability because sexual harassment rarely, if ever, is among the official duties of a supervisor"). But cf. Martin v. Cavalier Hotel Corp., 48 F. 3d 1343, 1351–1352 (CA4 1995) (holding employer vicariously liable in part based on finding that the supervisor's rape of employee was within the scope of employment); Kauffman v. Allied Signal, Inc., 970 F. 2d 178, 184 (CA6) (holding that a supervisor's harassment was within the scope of his employment, but nevertheless requiring the victim to show that the employer failed to respond adequately when it learned of the harassment), cert. denied, 506 U.S. 1041 (1992). In so doing, the courts have emphasized that harassment consisting of unwelcome remarks and touching is motivated solely by individual desires and serves no purpose of the employer. For this reason, courts have likened hostile environment sexual harassment to the classic "frolic and detour" for which an employer has no vicarious liability.

These cases ostensibly stand in some tension with others arising outside Title VII, where the scope of employment has been defined broadly enough to hold employers vicariously liable for intentional torts that were in no sense inspired by any purpose to serve the employer. In Ira S. Bushey & Sons, Inc. v. United States, 398 F. 2d 167 (1968), for example, the Second Circuit charged the Government with vicarious liability for the depredation of a drunken sailor returning to his ship after a night's carouse, who inexplicably opened valves that flooded a dry dock, damaging both the drydock and the ship. Judge Friendly acknowledged that the sailor's conduct was not remotely motivated by a purpose to serve his employer, but relied on the "deeply rooted sentiment that a business enterprise cannot justly disclaim responsibility for accidents which may fairly be said to be characteristic of its activities," and imposed vicarious liability on the ground that the sailor's conduct "was not so 'unforeseeable' as to make it unfair to charge the Government with responsibility." Id., at 171. Other examples of an expansive sense of scope of employment are readily found, see, e.g., Leonbruno v. Champlain Silk Mills, 229 N. Y. 470, 128 N. E. 711 (1920) (opinion of Cardozo, J.) (employer was liable under worker's compensation statute for eye injury sustained when employee threw an apple at another; the accident arose "in the course of employment" because such horseplay should be expected); Carr v. Wm. C. Crowell Co., 28 Cal. 2d 652, 171 P. 2d 5 (1946) (employer liable for actions of carpenter who attacked a co-employee with a hammer). Courts, in fact, have treated scope of employment generously enough to include sexual assaults. See, e.g., Primeaux v. United States, 102 F. 3d 1458, 1462–1463 (CA8 1996) (federal police officer on limited duty sexually assaulted stranded motorist); Mary M. v. Los Angeles, 54 Cal. 3d 202, 216– 221, 814 P. 2d 1341, 1349–1352 (1991) (en banc) (police officer raped motorist after placing her under arrest); Doe v. Samaritan Counseling Ctr., 791 P. 2d 344, 348–349 (Alaska 1990) (therapist had sexual relations with patient); Turner v. State, 494 So. 2d 1291, 1296 (La. App. 1986) (National Guard recruiting officer committed sexual battery during sham physical examinations); Lyon v. Carey, 533 F. 2d 649, 655 (CADC 1976) (furniture deliveryman raped recipient of furniture); Samuels v. Southern Baptist Hospital, 594 So. 2d 571, 574 (La. App. 1992) (nursing assistant raped patient).[2]

The rationales for these decisions have varied, with some courts echoing Bushey in explaining that the employees's acts were foreseeable and that the employer should in fairness bear the resulting costs of doing business, see, e.g., Mary M., supra, at 218, 814 P. 2d., at 1350, and others finding that the employee's sexual misconduct arose from or was in some way related to the employee's essential duties. See, e.g., Samuels, supra, at 574 (tortious conduct was "reasonably incidental" to the performance of the nursing assistant's duties in caring for a "helpless" patient in a "locked environment").

An assignment to reconcile the run of the Title VII cases with those just cited would be a taxing one. Here it is enough to recognize that their disparate results do not necessarily reflect wildly varying terms of the particular employment contracts involved, but represent differing judgments about the desirability of holding an employer liable for his subordinates' wayward behavior. In the instances in which there is a genuine question about the employer's responsibility for harmful conduct he did not in fact authorize, a holding that the conduct falls within the scope of employment ultimately expresses a conclusion not of fact but of law. As one eminent authority has observed, the "highly indefinite phrase" is "devoid of meaning in itself" and is "obviously no more than a bare formula to cover the unordered and unauthorized acts of the servant for which it is found to be expedient to charge the master with liability, as well as to exclude other acts for which it is not." W. Keeton, D. Dobbs, R. Keeton, & D. Owen, Prosser and Keeton on the Law of Torts 502 (5th ed. 1984); see also Seavey, Speculations as to "Respondeat Superior," in Studies in Agency 129, 155 (1949) ("The liability of a master to a third person for the torts of a servant has been widely extended by aid of the elastic phrase 'scope of the employment' which may be used to include all which the court wishes to put into it"). Older cases, for example, treated smoking by an employee during working hours as an act outside the scope of employment, but more recently courts have generally held smoking on the job to fall within the scope. Prosser & Keeton, supra, at 504, and n. 23. It is not that employers formerly did not authorize smoking but have now begun to do so, or that employees previously smoked for their own purposes but now do so to serve the employer. We simply understand smoking differently now and have revised the old judgments about what ought to be done about it.

The proper analysis here, then, calls not for a mechanical application of indefinite and malleable factors set forth in the Restatement, see, e.g., §§ 219, 228, 229, but rather an enquiry into the reasons that would support a conclusion that harassing behavior ought to be held within the scope of a supervisor's employment, and the reasons for the opposite view. The Restatement itself points to such an approach, as in the commentary that the "ultimate question" in determining the scope of employment is "whether or not it is just that the loss resulting from the servant's acts should be considered as one of the normal risks to be borne by the business in which the servant is employed." Id., § 229, Comment a. See generally Taber v. Maine, 67 F. 3d

1029, 1037 (CA2 1995) ("As the leading Torts treatise has put it, 'the integrating principle' of respondeat superior is 'that the employer should be liable for those faults that may be fairly regarded as risks of his business, whether they are committed in furthering it or not'") (quoting 5 F. Harper, F. James & O. Gray, Law of Torts § 26.8, pp. 40–41 (2d ed. 1986).

In the case before us, a justification for holding the offensive behavior within the scope of Terry's and Silverman's employment was well put in Judge Barkett's dissent: "[A] pervasively hostile work environment of sexual harassment is never (one would hope) authorized, but the supervisor is clearly charged with maintaining a productive, safe work environment. The supervisor directs and controls the conduct of the employees, and the manner of doing so may inure to the employer's benefit or detriment, including subjecting the employer to Title VII liability." 111 F. 3d, at 1542 (opinion dissenting in part and concurring in part). It is by now well recognized that hostile environment sexual harassment by supervisors (and, for that matter, co-employees) is a persistent problem in the workplace. See Lindemann & Kadue 4–5 (discussing studies showing prevalence of sexual harassment); Ellerth, 123 F. 3d, at 511 (Posner, C. J., concurring and dissenting) ("[E]veryone knows by now that sexual harassment is a common problem in the American workplace"). An employer can, in a general sense, reasonably anticipate the possibility of such conduct occurring in its workplace, and one might justify the assignment of the burden of the untoward behavior to the employer as one of the costs of doing business, to be charged to the enterprise rather than the victim. As noted, supra, at 17–18, developments like this occur from time to time in the law of agency.

Two things counsel us to draw the contrary conclusion. First, there is no reason to suppose that Congress wished courts to ignore the traditional distinction between acts falling within the scope and acts amounting to what the older law called frolics or detours from the course of employment. Such a distinction can readily be applied to the spectrum of possible harassing conduct by supervisors, as the following examples show. First, a supervisor might discriminate racially in job assignments in order to placate the prejudice pervasive in the labor force. Instances of this variety of the heckler's veto would be consciously intended to further the employer's interests by preserving peace in the workplace. Next, supervisors might reprimand male employees for workplace failings with banter, but respond to women's shortcomings in harsh or vulgar terms. A third example might be the supervisor who, as here, expresses his sexual interests in ways having no apparent object whatever of serving an interest of the employer. If a line is to be drawn between scope and frolic, it would lie between the first two examples and the third, and it thus makes sense in terms of traditional agency law to analyze the scope issue, in cases like the third example, just as most federal courts addressing that issue have done, classifying the harassment as beyond the scope of employment.

The second reason goes to an even broader unanimity of views among the holdings of District Courts and Courts of Appeals thus far. Those courts have held not only that the sort of harassment at issue here was outside the scope of supervisors' authority, but, by uniformly judging employer liability for co-worker harassment under a negligence standard, they have also implicitly treated such harassment as outside the scope of common employees' duties as well. See Blankenship v. Parke Care Centers, Inc., 123 F. 3d 868, 872–873 (CA6 1997), cert. denied, 522 U.S. ___ (1998); Fleming v. Boeing Co., 120 F. 3d 242, 246 (CA11 1997); Perry v. Ethan Allen, Inc., 115 F. 3d 143, 149 (CA2 1997); Yamaguchi v. United States Dept. of Air Force, 109 F. 3d 1475, 1483 (CA9 1997); Varner v. National Super Markets, Inc., 94 F. 3d 1209, 1213 (CA8 1996), cert. denied, 519 U.S. ___ (1997); McKenzie v. Illinois Dept. of Transp., 92 F. 3d 473, 480 (CA7 1996); Andrade, 88 F. 3d, at 261; Waymire v. Harris County, 86 F. 3d 424, 428–429 (CA5 1996); Hirase-Doi v. U.S. West Communications, Inc., 61 F. 3d 777, 783 (CA10 1995); Andrews v. Philadelphia, 895 F.2d 1469, 1486 (CA3 1990); cf. Morrison v. Carleton Woolen Mills, Inc., 108 F. 3d 429, 438 (CA1 1997) (applying "knew or should have known" standard to claims of environmental harassment by a supervisor); see also 29 CFR § 1604.11(d) (1997) (employer is liable for co-worker harassment if it "knows or should have known of the conduct, unless it can show that it took immediate and appropriate corrective action"); 3 L. Larson & A. Larson, Employment Discrimination § 46.07[4][a], p. 46-101 (2d ed. 1998) (courts "uniformly" apply EEOC rule; "[i]t is not a controversial area"). If, indeed, the cases did not rest, at least implicitly, on the notion that such harassment falls outside the scope of employment, their liability issues would have turned simply on the application of the scope-of-employment rule. Cf. Hunter v. Allis-Chalmers, Inc., 797 F. 2d 1417, 1422 (CA7 1986) (noting that employer will not usually be liable under respondeat superior for employee's racial harassment because it "would be the rare case where racial harassment . . . could be thought by the author of the harassment to help the employer's business").

It is quite unlikely that these cases would escape efforts to render them obsolete if we were to hold that supervisors who engage in discriminatory harassment are necessarily acting within the scope of their employment. The rationale for placing harassment within the scope of supervisory authority would be the fairness of requiring the employer to bear the burden of foreseeable social behavior, and the same rationale would apply when the behavior was that of co-employees. The employer generally benefits just as obviously from the work of common employees as from the work of supervisors; they simply have different jobs to do, all aimed at the success of the enterprise. As between an innocent employer and an innocent employee, if we use scope-of-employment reasoning to require the employer to bear the cost of an actionably hostile workplace created by one class of employees (i.e., supervisors), it could appear just as appropriate to do the same when the environment was created by another class (i.e., co-workers).

The answer to this argument might well be to point out that the scope of supervisory employment may be treated separately by recognizing that supervisors have special authority enhancing their capacity to harass, and that the employer can guard against their misbehavior more easily because their numbers are by definition fewer than the numbers of regular employees. But this answer happens to implicate an entirely separate category of agency law (to be considered in the next section), which imposes vicarious liability on employers for tortious acts committed by use of particular authority conferred as an element of an employee's agency relationship with the employer. Since the virtue of categorical clarity is obvious, it is better to reject reliance on misuse of supervisory authority (without more) as irrelevant to scope-of-employment analysis.

2

The Court of Appeals also rejected vicarious liability on the part of the City insofar as it might rest on the concluding principle set forth in § 219(2)(d) of the Restatement, that an employer "is not subject to liability for the torts of his servants acting outside the scope of their employment unless . . . the servant purported to act or speak on behalf of the principal and there was reliance on apparent authority, or he was aided in accomplishing the tort by the existence of the agency relation." Faragher points to several ways in which the agency relationship aided Terry and Silverman in carrying out their harassment. She argues that in general offending supervisors can abuse their authority to keep subordinates in their presence while they make offensive statements, and that they implicitly threaten to misuse their supervisory powers to deter any resistance or complaint. Thus, she maintains that power conferred on Terry and Silverman by the City enabled them to act for so long without provoking defiance or complaint.

The City, however, contends that 219(2)(d) has no application here. It argues that the second qualification of the subsection, referring to a servant "aided in accomplishing the tort by the existence of the agency relation," merely "refines" the one preceding it, which holds the employer vicariously liable for its servant's abuse of apparent authority. Brief for Respondent 30–31, and n. 24. But this narrow reading is untenable; it would render the second qualification of § 219(2)(d) almost entirely superfluous (and would seem to ask us to shut our eyes to the potential effects of supervisory authority, even when not explicitly invoked). The illustrations accompanying this subsection make clear that it covers not only cases involving the abuse of apparent authority, but also to cases in which tortious conduct is made possible or facilitated by the existence of the actual agency relationship. See Restatement § 219 Comment e (noting employer liability where "the servant may be able to cause harm because of his position as agent, as where a telegraph operator sends false messages purporting to come from third persons" and where the manager who operates a store "for an undisclosed principal is enabled to cheat the customers because of his position"); § 247, Illustration 1 (noting a newspaper's liability for a libelous editorial published by an editor acting for his own purposes).

We therefore agree with Faragher that in implementing Title VII it makes sense to hold an employer vicariously liable for some tortious conduct of a supervisor made possible by abuse of his supervisory authority, and that the aided-by-agency-relation principle embodied in § 219(2)(d) of the Restatement provides an appropriate starting point for determining liability for the kind of harassment presented here.[3]

Several courts, indeed, have noted what Faragher has argued, that there is a sense in which a harassing supervisor is always assisted in his misconduct by the supervisory relationship. See, e.g., Rodgers v. Western-Southern Life Ins. Co., 12 F. 3d 668, 675 (CA7 1993); Taylor v. Metzger, 152 N. J. 490, 505, 706 A. 2d 685, 692 (1998) (emphasizing that a supervisor's conduct may have a greater impact than that of colleagues at the same level); cf. Torres, 116 F. 3d, at 631. See also White v. Monsanto Co., 585 So. 2d 1205, 1209–1210 (La. 1991) (a supervisor's harassment of a subordinate is more apt to rise to the level of intentional infliction of emotional distress than comparable harassment by a co-employee); Contreras v. Crown Zellerbach Corp., 88 Wash. 2d 735, 740, 565 P. 2d 1173, 1176 (1977) (same); Alcorn v. Anbro Engineering, Inc., 2 Cal. 3d 493, 498–499, and n. 2, 468 P. 2d 216, 218–219, and n. 2 (1970) (same). The agency relationship affords contact with an employee subjected to a supervisor's sexual harassment, and the victim may well be reluctant to accept the risks of blowing the whistle on a superior. When a person with supervisory authority discriminates in the terms and conditions of subordinates' employment, his actions necessarily draw upon his superior position over the people who report to him, or those under them, whereas an employee generally cannot check a supervisor's abusive conduct the same way that she might deal with abuse from a co-worker. When a fellow employee harasses, the victim can walk away or tell the offender where to go, but it may be difficult to offer such responses to a supervisor, whose "power to supervise—[which may be] to hire and fire, and to set work schedules and pay rates—does not disappear . . . when he chooses to harass through insults and offensive gestures rather than directly with threats of firing or promises of promotion." Estrich, Sex at Work, 43 Stan. L. Rev. 813, 854 (1991). Recognition of employer liability when discriminatory misuse of supervisory authority alters the terms and conditions of a victim's employment is underscored by the fact that the employer has a greater opportunity to guard against misconduct by supervisors than by common workers; employers have greater opportunity and incentive to screen them, train them, and monitor their performance.

In sum, there are good reasons for vicarious liability for misuse of supervisory authority. That rationale must, however, satisfy one more condition. We are not entitled to recognize this theory under Title VII unless we can square it with Meritor's holding that an employer is not "automatically" liable for harassment by a supervisor who creates the requisite degree of discrimination,[4]

and there is obviously some tension between that holding and the position that a supervisor's misconduct aided by supervisory authority subjects the employer to liability vicariously; if the "aid" may be the unspoken suggestion of retaliation by misuse of supervisory authority, the risk of automatic liability is high. To counter it, we think there are two basic alternatives, one being to require proof of some affirmative invocation of that authority by the harassing supervisor, the other to recognize an affirmative defense to liability in some circumstances, even when a supervisor has created the actionable environment.

There is certainly some authority for requiring active or affirmative, as distinct from passive or implicit, misuse of supervisory authority before liability may be imputed. That is the way some courts have viewed the familiar cases holding the employer liable for discriminatory employment action with tangible consequences, like firing and demotion. See supra, at 11–12. And we have already noted some examples of liability provided by the Restatement itself, which suggests that an affirmative misuse of power might be required. See supra, at 23–24 (telegraph operator sends false messages, a store manager cheats customers, editor publishes libelous editorial).

But neat examples illustrating the line between the affirmative and merely implicit uses of power are not easy to come by in considering management behavior. Supervisors do not make speeches threatening sanctions whenever they make requests in the legitimate exercise of managerial authority, and yet every subordinate employee knows the sanctions exist; this is the reason that courts have consistently held that acts of supervisors have greater power to alter the environment than acts of coemployees generally, see supra, at 24–25. How far from the course of ostensible supervisory behavior would a company officer have to step before his orders would not reasonably be seen as actively using authority? Judgment calls would often be close, the results would often seem disparate even if not demonstrably contradictory, and the temptation to litigate would be hard to resist. We think plaintiffs and defendants alike would be poorly served by an active-use rule.

The other basic alternative to automatic liability would avoid this particular temptation to litigate, but allow an employer to show as an affirmative defense to liability that the employer had exercised reasonable care to avoid harassment and to eliminate it when it might occur, and that the complaining employee had failed to act with like reasonable care to take advantage of the employer's safeguards and otherwise to prevent harm that could have been avoided. This composite defense would, we think, implement the statute sensibly, for reasons that are not hard to fathom.

Although Title VII seeks "to make persons whole for injuries suffered on account of unlawful employment discrimination," Albemarle Paper Co. v. Moody, 422 U.S. 405, 418 (1975), its "primary objective," like that of any statute meant to influence primary conduct, is not to provide redress but to avoid harm. Id., at 417. As long ago as 1980, the Equal Employment Opportunity Commission (EEOC), charged with the enforcement of Title VII, 42 U.S.C. § 2000e-4, adopted regulations advising employers to "take all steps necessary to prevent sexual harassment from occurring, such as . . . informing employees of their right to raise and how to raise the issue of harassment." 29 CFR § 1604.11(f) (1997), and in 1990 the Commission issued a policy statement enjoining employers to establish a complaint procedure "designed to encourage victims of harassment to come forward [without requiring] a victim to complain first to the offending supervisor." EEOC Policy Guidance on Sexual Harassment, 8 FEP Manual 405:6699 (Mar. 19, 1990) (internal quotation marks omitted). It would therefore implement clear statutory policy and complement the Government's Title VII enforcement efforts to recognize the employer's affirmative obligation to prevent violations and give credit here to employers who make reasonable efforts to discharge their duty. Indeed, a theory of vicarious liability for misuse of supervisory power would be at odds with the statutory policy if it failed to provide employers with some such incentive.

The requirement to show that the employee has failed in a coordinate duty to avoid or mitigate harm reflects an equally obvious policy imported from the general theory of damages, that a victim has a duty "to use such means as are reasonable under the circumstances to avoid or minimize the damages" that result from violations of the statute. Ford Motor Co. v. EEOC, 458 U.S. 219, 231, n. 15 (1982) (quoting C. McCormick, Law of Damages 127 (1935) (internal quotation marks omitted). An employer may, for example, have provided a proven, effective mechanism for reporting and resolving complaints of sexual harassment, available to the employee without undue risk or expense. If the plaintiff unreasonably failed to avail herself of the employer's preventive or remedial apparatus, she should not recover damages that could have been avoided if she had done so. If the victim could have avoided harm, no liability should be found against the employer who had taken reasonable care, and if damages could reasonably have been mitigated no award against a liable employer should reward a plaintiff for what her own efforts could have avoided.

In order to accommodate the principle of vicarious liability for harm caused by misuse of supervisory authority, as well as Title VII's equally basic policies of encouraging forethought by employers and saving action by objecting employees, we adopt the following holding in this case and in Burlington Industries, Inc. v. Ellerth, ante, p. __, also decided today. An employer is subject to vicarious liability to a victimized employee for an actionable hostile environment created by a supervisor with immediate (or successively higher) authority over the employee. When no tangible employment action is taken, a defending employer may raise an affirmative defense to liability or damages, subject to proof by a preponderance of the evidence, see Fed. Rule. Civ. Proc. 8(c). The defense comprises two necessary elements: (a) that the employer exercised reasonable care to prevent and correct promptly any sexually harassing behavior, and (b) that the plaintiff employee unreasonably failed to take advantage of any preventive or corrective opportunities provided by the employer or to avoid harm otherwise. While proof that an employer had promulgated an antiharassment policy with complaint procedure is not necessary in every instance as a matter of law, the need for a stated policy suitable to the employment circumstances may appropriately be addressed in any case when litigating the first element of the defense. And while proof that an employee failed to fulfill the corresponding obligation of reasonable care to avoid harm is not limited to showing an unreasonable failure to use any complaint procedure provided by the employer, a demonstration of such failure will normally suffice to satisfy the employer's burden under the second element of the defense. No affirmative defense is available, however, when the supervisor's harassment culminates in a tangible employment action, such as discharge, demotion, or undesirable reassignment. See Burlington, ante, at 17.

Applying these rules here, we believe that the judgment of the Court of Appeals must be reversed. The District Court found that the degree of hostility in the work environment rose to the actionable level and was attributable to Silverman and Terry. It is undisputed that these supervisors "were granted virtually unchecked authority" over their subordinates, "directly controll[ing] and supervis[ing] all aspects of [Faragher's] day-to-day activities." 111 F. 3d, at, 1544 (Barkett, J., dissenting in part and concurring in part). It is also clear that Faragher and her colleagues were "completely isolated from the City's higher management." Ibid. The City did not seek review of these findings.

While the City would have an opportunity to raise an affirmative defense if there were any serious prospect of its presenting one, it appears from the record that any such avenue is closed. The District Court found that the City had entirely failed to disseminate its policy against sexual harassment among the beach employees and that its officials made no attempt to keep track of the conduct of supervisors like Terry and Silverman. The record also makes clear that the City's policy did not include any assurance that the harassing supervisors could be bypassed in registering complaints. App. 274. Under such circumstances, we hold as a matter of law that the City could not be found to have exercised reasonable care to prevent the supervisors' harassing conduct. Unlike the employer of a small workforce, who might expect that sufficient care to prevent tortious behavior could be exercised informally, those responsible for city operations could not reasonably have thought that precautions against hostile environments in any one of many departments in far-flung locations could be effective without communicating some formal policy against harassment, with a sensible complaint procedure.

We have drawn this conclusion without overlooking two possible grounds upon which the City might argue for the opportunity to litigate further. There is, first, the Court of Appeals's indulgent gloss on the relevant evidence: "There is some evidence that the City did not effectively disseminate among Marine Safety employees its sexual harassment policy." Id., at 1539, n. 11. But, in contrast to the Court of Appeals's characterization, the District Court made an explicit finding of a "complete failure on the part of the City to disseminate said policy among Marine Safety Section employees." 864 F. Supp., at 1560. The evidence supports the District Court's finding and there is no contrary claim before us.

The second possible ground for pursuing a defense was asserted by the City in its argument addressing the possibility of negligence liability in this case. It said that it should not be held liable for failing to promulgate an antiharassment policy, because there was no apparent duty to do so in the 1985–1990 period. The City purports to rest this argument on the position of the EEOC during the period mentioned, but it turns out that the record on this point is quite against the City's position. Although the EEOC issued regulations dealing with promulgating a statement of policy and providing a complaint mechanism in 1990, see supra, at 28, ever since 1980 its regulations have called for steps to prevent violations, such as informing employees of their rights and the means to assert them. Ibid. The City, after all, adopted an antiharassment policy in 1986.

The City points to nothing that might justify a conclusion by the District Court on remand that the City had exercised reasonable care. Nor is there any reason to remand for consideration of Faragher's efforts to mitigate her own damages, since the award to her was solely nominal.

3

The Court of Appeals also rejected the possibility that it could hold the City liable for the reason that it knew of the harassment vicariously through the knowledge of its supervisors. We have no occasion to consider whether this was error, however. We are satisfied that liability on the ground of vicarious knowledge could not be determined without further factfinding on remand, whereas the reversal necessary on the theory of supervisory harassment renders any remand for consideration of imputed knowledge entirely unjustifiable (as would be any consideration of negligence as an alternative to a theory of vicarious liability here).

III

The judgment of the Court of Appeals for the Eleventh Circuit is reversed, and the case is remanded for reinstatement of the judgment of the District Court.

It is so ordered.

Notes

1. Similarly, Courts of Appeals in sexual harassment cases have properly drawn on standards developed in cases involving racial harassment. See, e.g., Carrero v. New York City Housing Auth., 890 F. 2d 569, 577 (CA2 1989) (citing Lopez v. S. B. Thomas, Inc., 831 F. 2d 1184, 1189 (CA2 1987), a case of racial harassment, for the proposition that incidents of environmental sexual harassment "must be more than episodic; they must be sufficiently continuous and concerted in order to be deemed pervasive"). Although racial and sexual harassment will often take different forms, and standards may not be entirely interchangeable, we think there is good sense in seeking generally to harmonize the standards of what amounts to actionable harassment.

2. It bears noting that many courts in non–Title VII cases have held sexual assaults to fall outside the scope of employment. See Note, "Scope of Employment" Redefined: Holding Employers Vicariously Liable for Sexual Assaults Committed by their Employees, 76 Minn. L. Rev. 1513, 1521–1522, and nn. 33, 34 (1992) (collecting cases).

3. We say "starting point" because our obligation here is not to make a pronouncement of agency law in general or to transplant § 219(2)(d) into Title VII. Rather, it is to adapt agency concepts to the practical objectives of Title VII. As we said in Meritor Savings Bank, FSB v. Vinson, 477 U.S. 57, 72 (1986), "common-law principles may not be transferable in all their particulars to Title VII."

4. We are bound to honor Meritor on this point not merely because of the high value placed on stare decisis in statutory interpretation, supra, at 13–14, but for a further reason as well. With the amendments enacted by the Civil Rights Act of 1991, Congress both expanded the monetary relief available under Title VII to include compensatory and punitive damages, see § 102, 105 Stat. 1072, 42 U.S.C. § 1981a and modified the statutory grounds of several of our decisions, see § 101 et seq. The decision of Congress to leave Meritor intact is conspicuous. We thus have to assume that in expanding employers' potential liability under Title VII, Congress relied on our statements in Meritor about the limits of employer liability. To disregard those statements now (even if we were convinced of reasons for doing so) would be not only to disregard stare decisis in statutory interpretation, but to substitute our revised judgment about the proper allocation of the costs of harassment for Congress's considered decision on the subject.

U.S. Supreme Court

No. 97-282

BETH ANN FARAGHER, PETITIONER v. CITY OF BOCA RATON

ON WRIT OF CERTIORARI TO THE UNITED STATES COURT OF APPEALS FOR THE ELEVENTH CIRCUIT

[June 26, 1998]

JUSTICE THOMAS, with whom JUSTICE SCALIA joins, dissenting.

For the reasons given in my dissenting opinion in Burlington Industries v. Ellerth, ante, absent an adverse employment consequence, an employer cannot be held vicariously liable if a supervisor creates a hostile work environment. Petitioner suffered no adverse employment consequence; thus the Court of Appeals was correct to hold that the City is not vicariously liable for the conduct of Chief Terry and Lieutenant Silverman. Because the Court reverses this judgment, I dissent.

As for petitioner's negligence claim, the District Court made no finding as to the City's negligence, and the Court of Appeals did not directly consider the issue. I would therefore remand the case to the District Court for further proceedings on this question alone. I disagree with the Court's conclusion that merely because the City did not disseminate its sexual harassment policy, it should be liable as a matter of law. See ante, at 31.[1]

The City should be allowed to show either that: (1) there was a reasonably available avenue through which petitioner could have complained to a City official who supervised both Chief Terry and Lieutenant Silverman, see Brief for United States and EEOC as Amici Curiae in Meritor Savings Bank, FSB v. Vinson, O.T. 1985, No. 84-1979, p. 26,[2] or (2) it would not have learned of the harassment even if the policy had been distributed.[3]

Petitioner, as the plaintiff, would of course bear the burden of proving the City's negligence.

Notes

1. The harassment alleged in this case occurred intermittently over a 5-year period between 1985 and 1990; the District Court's factual findings do not indicate when in 1990 it ceased. It was only in March 1990 that the Equal Employment Opportunity Commission (EEOC) issued a "policy statement" "enjoining" employers to establish complaint procedures for sexual harassment. See ante, at 28. The 1980 Guideline on which the Court relies—because the EEOC has no substantive rulemaking authority under Title VII, the Court is inaccurate to refer to it as a "regulation," see ante, at 32,—was wholly precatory and as such cannot establish negligence per se. See 29 CFR § 1604.11(f) (1997) ("An employer should take all steps necessary to prevent sexual harassment from occurring.").

2. The City's Employment Handbook stated that employees with "complaints or grievances" could speak to the City's Personnel and Labor Relations Director about problems at work. See App. 280. The District Court found that the City's Personnel Director, Richard Bender, moved quickly to investigate the harassment charges against Terry and Silverman once they were brought to his attention. See App. to Pet. for Cert. 80a.

3. Even after petitioner read the City's sexual harassment policy in 1990, see App. 188, she did not file a charge with City officials. Instead, she filed suit against the City in 1992.

- [U.S.]
- [U.S.]
- [U.S.]

APPENDIX B

Burlington Industries, Inc. v. Ellerth

U.S. SUPREME COURT

Syllabus

BURLINGTON INDUSTRIES, INC. v. ELLERTH

CERTIORARI TO THE UNITED STATES COURT OF APPEALS
FOR THE SEVENTH CIRCUIT

No. 97-569.
Argued April 22, 1998–
Decided June 26, 1998

Respondent Kimberly Ellerth quit her job after 15 months as a salesperson in one of petitioner Burlington Industries' many divisions, allegedly because she had been subjected to constant sexual harassment by one of her supervisors, Ted Slowik. Slowik was a mid-level manager who had authority to hire and promote employees, subject to higher approval, but was not considered a policy-maker. Against a background of repeated boorish and offensive remarks and gestures allegedly made by Slowik, Ellerth places particular emphasis on three incidents where Slowik's comments could be construed as threats to deny her tangible job benefits. Ellerth refused all of Slowik's advances, yet suffered no tangible retaliation and was, in fact, promoted once. Moreover, she never informed anyone in authority about Slowik's conduct, despite knowing Burlington had a policy against sexual harassment. In filing this lawsuit, Ellerth alleged Burlington engaged in sexual harassment and forced her constructive discharge, in violation of Title VII of the Civil Rights Act of 1964, 42 U.S.C. § 2000e et seq. The District Court granted Burlington summary judgment. The Seventh Circuit en banc reversed in a decision that produced eight separate opinions and no consensus for a controlling rationale. Among other things, those opinions focused on whether Ellerth's claim could be categorized as one of quid pro quo harassment, and on whether the standard for an employer's liability on such a claim should be vicarious liability or negligence.

Held: Under Title VII, an employee who refuses the unwelcome and threatening sexual advances of a supervisor, yet suffers no adverse, tangible job consequences, may recover against the employer without showing the employer is negligent or otherwise at fault for the supervisor's actions, but the employer may interpose an affirmative defense. Pp. 6–21.

(a) The Court assumes an important premise yet to be established: a trier of fact could find in Slowik's remarks numerous threats to retaliate against Ellerth if she denied some sexual liberties. The threats, however, were not carried out. Cases based on carried-out threats are referred to often as "quid pro quo" cases, as distinct from bothersome attentions or sexual remarks sufficient to create a "hostile work environment." Those two terms do not appear in Title VII, which forbids only "discriminat[ion] against any individual with respect to his ... terms [or] conditions ... of employment, because of ... sex." § 2000e-2(a)(1). In Meritor Savings Bank, FSB v. Vinson, 477 U.S. 57, 65, this Court distinguished between the two concepts, saying both are cognizable under Title VII, though a hostile environment claim requires harassment that is severe or pervasive. Meritor did not discuss the distinction for its bearing upon an employer's liability for discrimination, but held, with no further specifics, that agency principles controlled on this point. Id., at 72. Nevertheless, in Meritor's wake, Courts of Appeals held that, if the plaintiff established a quid pro quo claim, the employer was subject to vicarious liability. This rule encouraged Title VII plaintiffs to state their claims in quid pro quo terms, which in turn put expansive pressure on the definition. For example, the question presented here is phrased as whether Ellerth can state a quid pro quo claim, but the issue of real concern to the parties is whether Burlington has vicarious liability, rather than liability limited to its own negligence. This Court nonetheless believes the two terms are of limited utility. To the extent they illustrate the distinction between cases involving a carried-out threat and offensive conduct in general, they are relevant when there is a threshold question whether a plaintiff can prove discrimination. Hence, Ellerth's claim involves only unfulfilled threats, so it is a hostile work environment claim requiring a showing of severe or pervasive conduct. This Court accepts the District Court's finding that Ellerth made such a showing. When discrimination is thus proved, the factors discussed below, not the categories quid pro quo and hostile work environment, control on the issue of vicarious liability. Pp. 6–9.

(b) In deciding whether an employer has vicarious liability in a case such as this, the Court turns to agency law principles, for Title VII defines the term "employer" to include "agents." § 2000e(b). Given this express direction, the Court concludes a uniform and predictable standard must be established as a matter of federal law. The Court relies on the general common law of agency, rather than on the law of any particular State. Community for Creative NonViolence v. Reid, 490 U.S. 730, 740. The Restatement (Second) of Agency (hereinafter Restatement) is a useful beginning point, although common-law principles may not be wholly transferable to Title VII. See Meritor, supra, at 72. Pp. 9–10.

(c) A master is subject to liability for the torts of his servants committed while acting in the scope of their employment. Restatement § 219(1). Although such torts generally may be either negligent or intentional, sexual harassment under Title VII presupposes intentional conduct. An intentional tort is within the scope of employment when actuated, at least in part, by a purpose to serve the employer. Id., §§ 228(1)(c), 230. Courts of Appeals have held, however, a supervisor acting out of gender-based animus or a desire to fulfill sexual urges may be actuated by personal motives unrelated and even antithetical to the employer's objectives. Thus, the general rule is that sexual harassment by a supervisor is not conduct within the scope of employment. Pp. 10–12.

(d) However, scope of employment is not the only basis for employer liability under agency principles. An employer is subject to liability for the torts of its employees acting outside the scope of their employment when, inter alia, the employer itself was negligent or reckless, Restatement § 219(2)(b), or the employee purported to act or to speak on behalf of the employer and there was reliance upon apparent authority, or he was aided in accomplishing the tort by the existence of the agency relation, id., § 219(2)(d). An employer is negligent, and therefore subject to liability under § 219(2)(b), if it knew or should have known about sexual harassment and failed to stop it. Negligence sets a minimum standard for Title VII liability; but Ellerth seeks to invoke the more stringent standard of vicarious liability. Section 219(2)(d) makes an employer vicariously liable for sexual harassment by an employee who uses apparent authority (the apparent authority standard), or who was "aided in accomplishing the tort by the existence of the agency relation" (the aided in the agency relation standard). Pp. 12–14.

(e) As a general rule, apparent authority is relevant where the agent purports to exercise a power which he or she does not have, as distinct from threatening to misuse actual power. Compare Restatement §§ 6 and 8. Because supervisory harassment cases involve misuse of actual power, not the false impression of its existence, apparent authority analysis is inappropriate. When a party seeks to impose vicarious liability based on an agent's misuse of delegated authority, the Restatement's aided in the agency relation rule provides the appropriate analysis. P. 14.

(f) That rule requires the existence of something more than the employment relation itself because, in a sense, most workplace tortfeasors, whether supervisors or co-workers, are aided in accomplishing their tortious objective by the employment relation: Proximity and regular contact afford a captive pool of potential victims. Such an additional aid exists when a supervisor subjects a subordinate to a significant, tangible employment action, i.e., a significant change in employment status, such as discharge, demotion, or undesirable reassignment. Every Federal Court of Appeals to have considered the question has correctly found vicarious liability in that circumstance. This Court imports the significant, tangible employment action concept for resolution of the vicarious liability issue considered here. An employer is therefore subject to vicarious liability for such actions. However, where, as here, there is no tangible employment action, it is not obvious the agency relationship aids in commission of the tort. Moreover, Meritor holds that agency principles constrain the imposition of employer liability for supervisor harassment. Limiting employer liability is also consistent with Title VII's purpose to the extent it would encourage the creation and use of anti-harassment policies and grievance procedures. Thus, in order to accommodate the agency principle of vicarious liability for harm caused by misuse of supervisory authority, as well as Title VII's equally basic policies of encouraging forethought by employers and saving action by objecting employees, the Court adopts, in this case and in Faragher v. Boca Raton, post, p. ___, the following holding: An employer is subject to vicarious liability to a victimized employee for an actionable hostile environment created by a supervisor with immediate (or successively higher) authority over the employee. When no tangible employment action is taken, a defending employer may raise an affirmative defense to liability or damages, subject to proof by a preponderance of the evidence, see Fed. Rule. Civ. Proc. 8(c). The defense comprises two necessary elements: (a) that the employer exercised reasonable care to prevent and correct promptly any sexually harassing behavior, and (b) that the plaintiff employee unreasonably failed to take advantage of any preventive or corrective opportunities provided by the employer or to avoid harm otherwise. While proof that an employer had promulgated an anti-harassment policy with a complaint procedure is not necessary in every instance as a matter of law, the need for a stated policy suitable to the employment circumstances may appropriately be addressed in any case when litigating the first element of the defense. And while proof that an employee failed to fulfill the corresponding obligation of reasonable care to avoid harm is not limited to showing any unreasonable failure to use any complaint procedure provided by the employer, a demonstration of such failure will normally suffice to satisfy the employer's burden under the second element of the defense. No affirmative defense is available, however, when the supervisor's harassment culminates in a tangible employment action. Pp. 15–20. (g) Given the Court's explanation that the labels quid pro quo and hostile work environment are not controlling for employer-liability purposes, Ellerth should have an adequate opportunity on remand to prove she has a claim which would result in vicarious liability. Although she has not alleged she suffered a tangible employment action at Slowik's hands, which would deprive Burlington of the affirmative defense, this is not dispositive. In light of the Court's decision, Burlington is still subject to vicarious liability for Slowik's activity, but should have an opportunity to assert and prove the affirmative defense. Pp. 20–21.

123 F. 3d 490, affirmed.

KENNEDY, J., delivered the opinion of the Court, in which REHNQUIST, C. J., and STEVENS, O'CONNOR, SOUTER, and BREYER, JJ., joined. GINSBURG, J., filed an opinion concurring in the judgment. THOMAS, J., filed a dissenting opinion, in which SCALIA, J., joined.

U.S. Supreme Court

No. 97-569

BURLINGTON INDUSTRIES, INC., PETITIONER
v. KIMBERLY B. ELLERTH

ON WRIT OF CERTIORARI TO THE UNITED STATES
COURT OF APPEALS FOR THE SEVENTH CIRCUIT

[June 26, 1998]

JUSTICE KENNEDY delivered the opinion of the Court.

We decide whether, under Title VII of the Civil Rights Act of 1964, 78 Stat. 253, as amended, 42 U.S.C. § 2000e et seq., an employee who refuses the unwelcome and threatening sexual advances of a supervisor, yet suffers no adverse, tangible job consequences, can recover against the employer without showing the employer is negligent or otherwise at fault for the supervisor's actions.

I

Summary judgment was granted for the employer, so we must take the facts alleged by the employee to be true. United States v. Diebold, Inc. 369 U.S. 654, 655 (1962) (per curiam). The employer is Burlington Industries, the petitioner. The employee is Kimberly Ellerth, the respondent. From March 1993 until May 1994, Ellerth worked as a salesperson in one of Burlington's divisions in Chicago, Illinois. During her employment, she alleges, she was subjected to constant sexual harassment by her supervisor, one Ted Slowik.

In the hierarchy of Burlington's management structure, Slowik was a mid-level manager. Burlington has eight divisions, employing more than 22,000 people in some 50 plants around the United States. Slowik was a vice president in one of five business units within one of the divisions. He had authority to make hiring and promotion decisions subject to the approval of his supervisor, who signed the paperwork. See 912 F. Supp. 1101, 1119, n. 14 (ND Ill. 1996). According to Slowik's supervisor, his position was "not considered an upper-level management position," and he was "not amongst the decision-making or policy-making hierarchy." Ibid. Slowik was not Ellerth's immediate supervisor. Ellerth worked in a two-person office in Chicago, and she answered to her office colleague, who in turn answered to Slowik in New York.

Against a background of repeated boorish and offensive remarks and gestures which Slowik allegedly made, Ellerth places particular emphasis on three alleged incidents where Slowik's comments could be construed as threats to deny her tangible job benefits. In the summer of 1993, while on a business trip, Slowik invited Ellerth to the hotel lounge, an invitation Ellerth felt compelled to accept because Slowik was her boss. App. 155. When Ellerth gave no encouragement to remarks Slowik made about her breasts, he told her to "loosen up" and warned, "[y]ou know, Kim, I could make your life very hard or very easy at Burlington." Id., at 156.

In March 1994, when Ellerth was being considered for a promotion, Slowik expressed reservations during the promotion interview because she was not "loose enough." Id., at 159. The comment was followed by his reaching over and rubbing her knee. Ibid. Ellerth did receive the promotion; but when Slowik called to announce it, he told Ellerth, "you're gonna be out there with men who work in factories, and they certainly like women with pretty butts/legs." Id., at 159–160.

In May 1994, Ellerth called Slowik, asking permission to insert a customer's logo into a fabric sample. Slowik responded, "I don't have time for you right now, Kim—unless you want to tell me what you're wearing." Id., at 78. Ellerth told Slowik she had to go and ended the call. Ibid. A day or two later, Ellerth called Slowik to ask permission again. This time he denied her request, but added something along the lines of, "are you wearing shorter skirts yet, Kim, because it would make your job a whole heck of a lot easier." Id., at 79.

A short time later, Ellerth's immediate supervisor cautioned her about returning telephone calls to customers in a prompt fashion. 912 F. Supp., at 1109. In response, Ellerth quit. She faxed a letter giving reasons unrelated to the alleged sexual harassment we have described. Ibid. About three weeks later, however, she sent a letter explaining she quit because of Slowik's behavior. Ibid.

During her tenure at Burlington, Ellerth did not inform anyone in authority about Slowik's conduct, despite knowing Burlington had a policy against sexual harassment. Ibid. In fact, she chose not to inform her immediate supervisor (not Slowik) because " 'it would be his duty as my supervisor to report any incidents of sexual harassment.' " Ibid. On one occasion, she told Slowik a comment he made was inappropriate. Ibid.

In October 1994, after receiving a right-to-sue letter from the Equal Employment Opportunity Commission (EEOC), Ellerth filed suit in the United States

District Court for the Northern District of Illinois, alleging Burlington engaged in sexual harassment and forced her constructive discharge, in violation of Title VII. The District Court granted summary judgment to Burlington. The Court found Slowik's behavior, as described by Ellerth, severe and pervasive enough to create a hostile work environment, but found Burlington neither knew nor should have known about the conduct. There was no triable issue of fact on the latter point, and the Court noted Ellerth had not used Burlington's internal complaint procedures. Id., at 1118. Although Ellerth's claim was framed as a hostile work environment complaint, the District Court observed there was a quid pro quo "component" to the hostile environment. Id., at 1121. Proceeding from the premise that an employer faces vicarious liability for quid pro quo harassment, the District Court thought it necessary to apply a negligence standard because the quid pro quo merely contributed to the hostile work environment. See id., at 1123. The District Court also dismissed Ellerth's constructive discharge claim.

The Court of Appeals en banc reversed in a decision which produced eight separate opinions and no consensus for a controlling rationale. The judges were able to agree on the problem they confronted: Vicarious liability, not failure to comply with a duty of care, was the essence of Ellerth's case against Burlington on appeal. The judges seemed to agree Ellerth could recover if Slowik's unfulfilled threats to deny her tangible job benefits was sufficient to impose vicarious liability on Burlington. Jansen v. Packing Corp. of America, 123 F. 3d 490, 494 (CA7 1997) (per curiam). With the exception of Judges Coffey and Easterbrook, the judges also agreed Ellerth's claim could be categorized as one of quid pro quo harassment, even though she had received the promotion and had suffered no other tangible retaliation. Ibid.

The consensus disintegrated on the standard for an employer's liability for such a claim. Six judges, Judges Flaum, Cummings, Bauer, Evans, Rovner, and Diane P. Wood, agreed the proper standard was vicarious liability, and so Ellerth could recover even though Burlington was not negligent. Ibid. They had different reasons for the conclusion. According to Judges Flaum, Cummings, Bauer, and Evans, whether a claim involves a quid pro quo determines whether vicarious liability applies; and they in turn defined quid pro quo to include a supervisor's threat to inflict a tangible job injury whether or not it was completed. Id., at 499. Judges Wood and Rovner interpreted agency principles to impose vicarious liability on employers for most claims of supervisor sexual harassment, even absent a quid pro quo. Id., at 565.

Although Judge Easterbrook did not think Ellerth had stated a quid pro quo claim, he would have followed the law of the controlling State to determine the employer's liability, and by this standard, the employer would be liable here. Id., at 552. In contrast, Judge Kanne said Ellerth had stated a quid pro quo claim, but negligence was the appropriate standard of liability when the quid pro quo involved threats only. Id., at 505.

Chief Judge Posner, joined by Judge Manion, disagreed. He asserted Ellerth could not recover against Burlington despite having stated a quid pro quo claim. According to Chief Judge Posner, an employer is subject to vicarious liability for "act[s] that significantly alter the terms or conditions of employment," or "company act[s]." Id., at 515. In the emergent terminology, an unfulfilled quid pro quo is a mere threat to do a company act rather than the act itself, and in these circumstances, an employer can be found liable for its negligence only. Ibid. Chief Judge Posner also found Ellerth failed to create a triable issue of fact as to Burlington's negligence. Id., at 517.

Judge Coffey rejected all of the above approaches because he favored a uniform standard of negligence in almost all sexual harassment cases. Id., at 518.

The disagreement revealed in the careful opinions of the judges of the Court of Appeals reflects the fact that Congress has left it to the courts to determine controlling agency law principles in a new and difficult area of federal law. We granted certiorari to assist in defining the relevant standards of employer liability. 522 U.S. ___ (1998).

II

At the outset, we assume an important proposition yet to be established before a trier of fact. It is a premise assumed as well, in explicit or implicit terms, in the various opinions by the judges of the Court of Appeals. The premise is: a trier of fact could find in Slowik's remarks numerous threats to retaliate against Ellerth if she denied some sexual liberties. The threats, however, were not carried out or fulfilled. Cases based on threats which are carried out are referred to often as quid pro quo cases, as distinct from bothersome attentions or sexual remarks that are sufficiently severe or pervasive to create a hostile work environment. The terms quid pro quo and hostile work environment are helpful, perhaps, in making a rough demarcation between cases in which threats are carried out and those where they are not or are absent altogether, but beyond this are of limited utility.

Section 703(a) of Title VII forbids

"an employer"(1) to fail or refuse to hire or to discharge any individual, or otherwise to discriminate against any individual with respect to his compensation, terms, conditions or privileges of employment, because of such individual's . . . sex."

42 U.S.C. § 2000e-2(a)(1).

"Quid pro quo" and "hostile work environment" do not appear in the statutory text. The terms appeared first in the academic literature, see C. MacKinnon, Sexual Harassment of Working Women (1979); found their way into decisions of the Courts of Appeals, see, e.g., Henson v. Dundee, 682 F. 2d 897, 909 (CA11 1982); and were mentioned in this Court's decision in Meritor Savings Bank, FSB v. Vinson, 477 U.S. 57 (1986). See generally E. Scalia, The Strange Career of Quid Pro Quo Sexual Harassment, 21 Harv. J. L. & Pub. Policy 307 (1998).

In Meritor, the terms served a specific and limited purpose. There we considered whether the conduct in question constituted discrimination in the terms or conditions of employment in violation of Title VII. We assumed, and with adequate reason, that if an employer demanded sexual favors from an employee in return for a job benefit, discrimination with respect to terms or conditions of employment was explicit. Less obvious was whether an employer's sexually demeaning behavior altered terms or conditions of employment in violation of Title VII. We distinguished between quid pro quo claims and hostile environment claims, see 477 U.S., at 65, and said both were cognizable under Title VII, though the latter requires harassment that is severe or pervasive. Ibid. The principal significance of the distinction is to instruct that Title VII is violated by either explicit or constructive alterations in the terms or conditions of employment and to explain the latter must be severe or pervasive. The distinction was not discussed for its bearing upon an employer's liability for an employee's discrimination. On this question Meritor held, with no further specifics, that agency principles controlled. Id., at 72.

Nevertheless, as use of the terms grew in the wake of Meritor, they acquired their own significance. The standard of employer responsibility turned on which type of harassment occurred. If the plaintiff established a quid pro quo claim, the Courts of Appeals held, the employer was subject to vicarious liability. See Davis v. Sioux City, 115 F. 3d 1365, 1367 (CA8 1997); Nichols v. Frank, 42 F. 3d 503, 513–514 (CA9 1994); Bouton v. BMW of North America, Inc., 29 F. 3d 103, 106–107 (CA3 1994); Sauers v. Salt Lake County, 1 F. 3d 1122, 1127 (CA10 1993); Kauffman v. Allied Signal, Inc., 970 F. 2d 178, 185–186 (CA6), cert. denied, 506 U.S. 1041 (1992); Steele v. Offshore Shipbuilding, Inc., 867 F. 2d 1311, 1316 (CA11 1989). The rule encouraged Title VII plaintiffs to state their claims as quid pro quo claims, which in turn put expansive pressure on the definition. The equivalence of the quid pro quo label and vicarious liability is illustrated by this case. The question presented on certiorari is whether Ellerth can state a claim of quid pro quo harassment, but the issue of real concern to the parties is whether Burlington has vicarious liability for Slowik's alleged misconduct, rather than liability limited to its own negligence. The question presented for certiorari asks:

> "Whether a claim of quid pro quo sexual harassment may be stated under Title VII. . . . where the plaintiff employee has neither submitted to the sexual advances of the alleged harasser nor suffered any tangible effects on the compensation, terms, conditions or privileges of employment as a consequence of a refusal to submit to those advances?"

Pet. for Cert. i.

We do not suggest the terms quid pro quo and hostile work environment are irrelevant to Title VII litigation. To the extent they illustrate the distinction between cases involving a threat which is carried out and offensive conduct in general, the terms are relevant when there is a threshold question whether a plaintiff can prove discrimination in violation of Title VII. When a plaintiff proves that a tangible employment action resulted from a refusal to submit to a supervisor's sexual demands, he or she establishes that the employment decision itself constitutes a change in the terms and conditions of employment that is actionable under Title VII. For any sexual harassment preceding the employment decision to be actionable, however, the conduct must be severe or pervasive. Because Ellerth's claim involves only unfulfilled threats, it should be categorized as a hostile work environment claim which requires a showing of severe or pervasive conduct. See Oncale v. Sundowner Offshore Services, Inc., 523 U.S. __, __ (1998) (slip op., at 6); Harris v. Forklift Systems, Inc., 510 U.S. 17, 21 (1993). For purposes of this case, we accept the District Court's finding that the alleged conduct was severe or pervasive. See supra, at 3. The case before us involves numerous alleged threats, and we express no opinion as to whether a single unfulfilled threat is sufficient to constitute discrimination in the terms or conditions of employment.

When we assume discrimination can be proved, however, the factors we discuss below, and not the categories quid pro quo and hostile work environment, will be controlling on the issue of vicarious liability. That is the question we must resolve.

III

We must decide, then, whether an employer has vicarious liability when a supervisor creates a hostile work environment by making explicit threats to alter a subordinate's terms or conditions of employment, based on sex, but does not fulfill the threat. We turn to principles of agency law, for the term "employer" is defined under Title VII to include "agents." 42 U.S.C. § 2000e(b); see Meritor, supra, at 72. In express terms, Congress has directed federal courts to interpret Title VII based on agency principles. Given such an explicit instruction, we conclude a uniform and predictable standard must be established as a matter of federal law. We rely "on the general common law of agency, rather than on the law of any particular State, to give meaning to these terms." Community for Creative Non-Violence v. Reid, 490 U.S. 730, 740 (1989). The resulting federal rule, based on a body of case law developed over time, is statutory interpretation pursuant to congressional direction. This is not federal common law in "the strictest sense, i.e., a rule of decision that amounts, not simply to an interpretation of a federal statute . . ., but, rather, to the judicial 'creation' of a special federal rule of decision." Atherton v. FDIC, 519 U.S. 213, 218 (1997). State court decisions, applying state employment discrimination law, may be instructive in applying general agency principles, but, it is interesting to note, in many cases their determinations of employer liability under state law rely in large part on federal court decisions under Title VII. E.g., Arizona v. Schallock, 189 Ariz. 250, 259, 941 P. 2d 1275, 1284 (1997); Lehmann v. Toys 'R' Us, Inc., 132 N.J. 587, 622, 626 A. 2d 445, 463 (1993); Thompson v. Berta Enterprises, Inc., 72 Wash. App. 531, 537–539, 864 P. 2d 983, 986–988 (1994).

As Meritor acknowledged, the Restatement (Second) of Agency (1957) (hereinafter Restatement), is a useful beginning point for a discussion of general agency principles. 477 U.S., at 72. Since our decision in Meritor, federal courts have explored agency principles, and we find useful instruction in their decisions, noting that "common-law principles may not be transferable in all their particulars to Title VII." Ibid. The EEOC has issued Guidelines governing sexual harassment claims under Title VII, but they provide little guidance on the issue of employer liability for supervisor harassment. See 29 CFR § 1604.11(c) (1997) (vicarious liability for supervisor harassment turns on "the particular employment relationship and the job functions performed by the individual").

A

Section 219(1) of the Restatement sets out a central principle of agency law:

> "A master is subject to liability for the torts of his servants committed while acting in the scope of their employment."

An employer may be liable for both negligent and intentional torts committed by an employee within the scope of his or her employment. Sexual harassment under Title VII presupposes intentional conduct. While early decisions absolved employers of liability for the intentional torts of their employees, the law now imposes liability where the employee's "purpose, however misguided, is wholly or in part to further the master's business." W. Keeton, D. Dobbs, R. Keeton, & D. Owen, Prosser and Keeton on Law of Torts § 70, p. 505 (5th ed. 1984) (hereinafter Prosser and Keeton on Torts). In applying scope of employment principles to intentional torts, however, it is accepted that "it is less likely that a willful tort will properly be held to be in the course of employment and that the liability of the master for such torts will naturally be more limited." F. Mechem, Outlines of the Law of Agency § 394, p. 266 (P. Mechem 4th ed., 1952). The Restatement defines conduct, including an intentional tort, to be within the scope of employment when "actuated, at least in part, by a purpose to serve the [employer]," even if it is forbidden by the employer. Restatement §§ 228(1)(c), 230. For example, when a salesperson lies to a customer to make a sale, the tortious conduct is within the scope of employment because it benefits the employer by increasing sales, even though it may violate the employer's policies. See Prosser and Keeton on Torts § 70, at 505–506.

As Courts of Appeals have recognized, a supervisor acting out of gender-based animus or a desire to fulfill sexual urges may not be actuated by a purpose to serve the employer. See, e.g., Harrison v. Eddy Potash, Inc., 112 F. 3d 1437, 1444 (CA10 1997), cert. pending, No. 97-232; Torres v. Pisano, 116 F. 3d 625, 634, n. 10 (CA2 1997). But see Kauffman v. Allied Signal, Inc., 970 F. 2d, at 184–185 (holding harassing supervisor acted within scope of employment, but employer was not liable because of its quick and effective remediation). The harassing supervisor often acts for personal motives, motives unrelated and even antithetical to the objectives of the employer. Cf. Mechem, supra, § 368 ("for the time being [the supervisor] is conspicuously and unmistakably seeking a personal end"); see also Restatement § 235, Illustration 2 (tort committed while "[a]cting purely from personal ill will" not within the scope of employment); § 235, Illustration 3 (tort committed in retaliation for failing to pay the employee a bribe not within the scope of employment). There are instances, of course, where a supervisor engages in unlawful discrimination with the purpose, mistaken or otherwise, to serve the employer. E.g., Sims v. Montgomery County Comm'n, 766 F. Supp. 1052, 1075 (MD Ala. 1990) (supervisor acting in scope of employment where employer has a policy of discouraging women from seeking advancement and "sexual harassment was simply a way of furthering that policy").

The concept of scope of employment has not always been construed to require a motive to serve the employer. E.g., Ira S. Bushey & Sons, Inc. v. United States, 398 F. 2d 167, 172 (CA2 1968). Federal courts have nonetheless found similar limitations on employer liability when applying the agency laws of the States under the Federal Tort Claims Act, which makes the Federal Government liable for torts committed by employees within the scope of employment. 28 U.S.C. § 1346(b); see, e.g., Jamison v. Wiley, 14 F. 3d 222, 237 (CA4 1994) (supervisor's unfair criticism of subordinate's work in retaliation for rejecting his sexual advances not within scope of employment); Wood v. United States, 995 F. 2d 1122, 1123 (CA1 1993) (Breyer, C. J.) (sexual harassment amounting to assault and battery "clearly outside the scope of employment"); see also 2 L. Jayson & R. Longstreth, Handling Federal Tort Claims § 9.07[4], p. 9-211 (1998).

The general rule is that sexual harassment by a supervisor is not conduct within the scope of employment.

B

Scope of employment does not define the only basis for employer liability under agency principles. In limited circumstances, agency principles impose liability on employers even where employees commit torts outside the scope of employment. The principles are set forth in the much-cited § 219(2) of the Restatement:

"(2)A master is not subject to liability for the torts of his servants acting outside the scope of their employment, unless: "(a) the master intended the conduct or the consequences, or "(b) the master was negligent or reckless, or "(c) the conduct violated a non-delegable duty of the master, or "(d) the servant purported to act or to speak on behalf of the principal and there was reliance upon apparent authority, or he was aided in accomplishing the tort by the existence of the agency relation."

See also § 219, Comment e (Section 219(2) "enumerates the situations in which a master may be liable for torts of servants acting solely for their own purposes and hence not in the scope of employment").

Subsection (a) addresses direct liability, where the employer acts with tortious intent, and indirect liability, where the agent's high rank in the company makes him or her the employer's alter ego. None of the parties contend Slowik's rank imputes liability under this principle. There is no contention, furthermore, that a nondelegable duty is involved. See § 219(2)(c). So, for our purposes here, subsections (a) and (c) can be put aside.

Subsections (b) and (d) are possible grounds for imposing employer liability on account of a supervisor's acts and must be considered. Under subsection (b), an employer is liable when the tort is attributable to the employer's own negligence. § 219(2)(b). Thus, although a supervisor's sexual harassment is outside the scope of employment because the conduct was for personal motives, an employer can be liable, nonetheless, where its own negligence is a cause of the harassment. An employer is negligent with respect to sexual harassment if it knew or should have known about the conduct and failed to stop it. Negligence sets a minimum standard for employer liability under Title VII; but Ellerth seeks to invoke the more stringent standard of vicarious liability.

Subsection 219(2)(d) concerns vicarious liability for intentional torts committed by an employee when the employee uses apparent authority (the apparent authority standard), or when the employee "was aided in accomplishing the tort by the existence of the agency relation" (the aided in the agency relation standard). Ibid. As other federal decisions have done in discussing vicarious liability for supervisor harassment, e.g., Henson v. Dundee, 682 F. 2d 897, 909 (CA11 1982), we begin with § 219(2)(d).

C

As a general rule, apparent authority is relevant where the agent purports to exercise a power which he or she does not have, as distinct from where the agent threatens to misuse actual power. Compare Restatement § 6 (defining "power") with § 8 (defining "apparent authority"). In the usual case, a supervisor's harassment involves misuse of actual power, not the false impression of its existence. Apparent authority analysis therefore is inappropriate in this context. If, in the unusual case, it is alleged there is a false impression that the actor was a supervisor, when he in fact was not, the victim's mistaken conclusion must be a reasonable one. Restatement § 8, Comment c ("Apparent authority exists only to the extent it is reasonable for the third person dealing with the agent to believe that the agent is authorized"). When a party seeks to impose vicarious liability based on an agent's misuse of delegated authority, the Restatement's aided in the agency relation rule, rather than the apparent authority rule, appears to be the appropriate form of analysis.

D

We turn to the aided in the agency relation standard. In a sense, most workplace tortfeasors are aided in accomplishing their tortious objective by the existence of the agency relation: Proximity and regular contact may afford a captive pool of potential victims. See Gary v. Long, 59 F. 3d 1391, 1397 (CADC 1995). Were this to satisfy the aided in the agency relation standard, an employer would be subject to vicarious liability not only for all supervisor harassment, but also for all co-worker harassment, a result enforced by neither the EEOC nor any court of appeals to have considered the issue. See, e.g., Blankenship v. Parke Care Centers, Inc., 123 F. 3d 868, 872 (CA6 1997), cert. denied, 522 U.S. __ (1998) (sex discrimination); McKenzie v. Illinois Dept. of Transp., 92 F. 3d 473, 480 (CA7 1996) (sex discrimination); Daniels v. Essex Group, Inc., 937 F. 2d 1264, 1273 (CA7 1991) (race discrimination); see also 29 CFR 1604.11(d) (1997) ("knows or should have known" standard of liability for cases of harassment between "fellow employees"). The aided in the agency relation standard, therefore, requires the existence of something more than the employment relation itself.

At the outset, we can identify a class of cases where, beyond question, more than the mere existence of the employment relation aids in commission of the harassment: when a supervisor takes a tangible employment action against the subordinate. Every Federal Court of Appeals to have considered the question has found vicarious liability when a discriminatory act results in a tangible employment action. See, e.g., Sauers v. Salt Lake County, 1 F. 3d 1122, 1127 (CA10 1993) (" 'If the plaintiff can show that she suffered an economic injury from her supervisor's actions, the employer becomes strictly liable without any further showing . . .' "). In Meritor, we acknowledged this consensus. See 477 U.S., at 70–71 ("[T]he courts have consistently held employers liable for the discriminatory discharges of employees by supervisory personnel, whether or not the employer knew, or should have known, or approved of the supervisor's actions"). Although few courts have elaborated how agency principles support this rule, we think it reflects a correct application of the aided in the agency relation standard.

In the context of this case, a tangible employment action would have taken the form of a denial of a raise or a promotion. The concept of a tangible employment action appears in numerous cases in the Courts of Appeals discussing claims involving race, age, and national origin discrimination, as well as sex discrimination. Without endorsing the specific results of those decisions, we think it prudent to import the concept of a tangible employment action for resolution of the vicarious liability issue we consider here. A tangible employment action constitutes a significant change in employment status, such as hiring, firing, failing to promote, reassignment with significantly different responsibilities, or a decision causing a significant change in benefits. Compare Crady v. Liberty Nat. Bank & Trust Co. of Ind., 993 F. 2d 132, 136 (CA7 1993) ("A materially adverse change might be indicated by a termination of employment, a demotion evidenced by a decrease in wage or salary, a less distinguished title, a material loss of benefits, significantly diminished material responsibilities, or other indices that might be unique to a particular situation"), with Flaherty v. Gas Research Institute, 31 F. 3d 451, 456 (CA7 1994) (a "bruised ego" is not enough); Kocsis v. Multi-Care Management, Inc., 97 F. 3d 876, 887 (CA6 1996) (demotion without change in pay, benefits, duties, or prestige insufficient) and Harlston v. McDonnell Douglas Corp., 37 F. 3d 379, 382 (CA8 1994) (reassignment to more inconvenient job insufficient).

When a supervisor makes a tangible employment decision, there is assurance the injury could not have been inflicted absent the agency relation. A tangible employment action in most cases inflicts direct economic harm. As a general proposition, only a supervisor, or other person acting with the authority of the company, can cause this sort of injury. A co-worker can break a co-worker's arm as easily as a supervisor, and anyone who has regular contact with an employee can inflict psychological injuries by his or her offensive conduct. See Gary, 59 F. 3d, at 1397; Henson, 682 F. 2d, at 910; Barnes v. Costle, 561 F. 2d 983, 996 (CADC 1977) (MacKinnon, J., concurring). But one co-worker (absent some elaborate scheme) cannot dock another's pay, nor can one co-worker demote another. Tangible employment actions fall within the special province of the supervisor. The supervisor has been empowered by the company as a distinct class of agent to make economic decisions affecting other employees under his or her control.

Tangible employment actions are the means by which the supervisor brings the official power of the enterprise to bear on subordinates. A tangible employment decision requires an official act of the enterprise, a company act. The decision in most cases is documented in official company records, and may be subject to review by higher level supervisors. E.g., Shager v. Upjohn Co., 913 F. 2d 398, 405 (CA7 1990) (noting that the supervisor did not fire plaintiff; rather, the Career Path Committee did, but the employer was still liable because the Committee functioned as the supervisor's "cat's-paw"). The supervisor often must obtain the imprimatur of the enterprise and use its internal processes. See Kotcher v. Rosa & Sullivan Appliance Center, Inc., 957 F. 2d 59, 62 (CA2 1992) ("From the perspective of the employee, the supervisor and the employer merge into a single entity").

For these reasons, a tangible employment action taken by the supervisor becomes for Title VII purposes the act of the employer. Whatever the exact contours of the aided in the agency relation standard, its requirements will always be met when a supervisor takes a tangible employment action against a subordinate. In that instance, it would be implausible to interpret agency principles to allow an employer to escape liability, as Meritor itself appeared to acknowledge. See, supra, at 15.

Whether the agency relation aids in commission of supervisor harassment which does not culminate in a tangible employment action is less obvious. Application of the standard is made difficult by its malleable terminology, which can be read to either expand or limit liability in the context of supervisor harassment. On the one hand, a supervisor's power and authority invests his or her harassing conduct with a particular threatening character, and in this sense, a supervisor always is aided by the agency relation. See Meritor, 477 U.S., at 77 (Marshall, J., concurring in judgment) ("[I]t is precisely because the supervisor is understood to be clothed with the employer's authority that he is able to impose unwelcome sexual conduct on subordinates"). On the other hand, there are acts of harassment a supervisor might commit which might be the same acts a co-employee would commit, and there may be some circumstances where the supervisor's status makes little difference.

It is this tension which, we think, has caused so much confusion among the Courts of Appeals which have sought to apply the aided in the agency relation standard to Title VII cases. The aided in the agency relation standard, however, is a developing feature of agency law, and we hesitate to render a definitive explanation of our understanding of the standard in an area where other important considerations must affect our judgment. In particular, we are bound by our holding in Meritor that agency principles constrain the imposition of vicarious liability in cases of supervisory harassment. See Meritor, supra, at 72 ("Congress' decision to define 'employer' to include any 'agent' of an employer, 42 U.S.C. § 2000e(b), surely evinces an intent to place some limits on the acts of employees for which employers under Title VII are to be held responsible"). Congress has not altered Meritor 's rule even though it has made significant amendments to Title VII in the interim. See Illinois Brick Co. v. Illinois, 431 U.S. 720, 736 (1977) ("[W]e must bear in mind that considerations of stare decisis weigh heavily in the area of statutory construction, where Congress is free to change this Court's interpretation of its legislation").

Although Meritor suggested the limitation on employer liability stemmed from agency principles, the Court acknowledged other considerations might be relevant as well. See, 477 U.S., at 72 ("common-law principles may not be transferable in all their particulars to Title VII"). For example, Title VII is designed to encourage the creation of antiharassment policies and effective grievance mechanisms. Were employer liability to depend in part on an employer's effort to create such procedures, it would effect Congress' intention to promote conciliation rather than litigation in the Title VII context, see EEOC v. Shell Oil Co., 466 U.S. 54, 77 (1984), and the EEOC's policy of encouraging the development of grievance procedures. See 29 CFR § 1604.11(f) (1997); EEOC Policy Guidance on Sexual Harassment, 8 BNA FEP Manual 405:6699 (Mar. 19, 1990). To the extent limiting employer liability could encourage employees to report harassing conduct before it becomes severe or pervasive, it would also serve Title VII's deterrent purpose. See McKennon v. Nashville Banner Publishing Co., 513 U.S. 352, 358 (1995). As we have observed, Title VII borrows from tort law the avoidable consequences doctrine, see Ford Motor Co. v. EEOC, 458 U.S. 219, 231, n. 15 (1982), and the considerations which animate that doctrine would also support the limitation of employer liability in certain circumstances.

In order to accommodate the agency principles of vicarious liability for harm caused by misuse of supervisory authority, as well as Title VII's equally basic policies of encouraging forethought by employers and saving action by objecting employees, we adopt the following holding in this case and in Faragher v. Boca Raton, post, also decided today. An employer is subject to vicarious liability to a victimized employee for an actionable hostile environment created by a supervisor with immediate (or successively higher) authority over the employee. When no tangible employment action is taken, a defending employer may raise an affirmative defense to liability or damages, subject to proof by a preponderance of the evidence, see Fed. Rule Civ. Proc. 8(c). The defense comprises two necessary elements: (a) that the employer exercised reasonable care to prevent and correct promptly any sexually harassing behavior, and (b) that the plaintiff employee unreasonably failed to take advantage of any preventive or corrective opportunities provided by the employer or to avoid harm otherwise. While proof that an employer had promulgated an anti-harassment policy with complaint procedure is not necessary in every instance as a matter of law, the need for a stated policy suitable to the employment circumstances may appropriately be addressed in any case when litigating the first element of the defense. And while proof that an employee failed to fulfill the corresponding obligation of reasonable care to avoid harm is not limited to showing any unreasonable failure to use any complaint procedure provided by the employer, a demonstration of such failure will normally suffice to satisfy the employer's burden under the second element of the defense. No affirmative defense is available, however, when the supervisor's harassment culminates in a tangible employment action, such as discharge, demotion, or undesirable reassignment.

IV

Relying on existing case law which held out the promise of vicarious liability for all quid pro quo claims, see supra, at 7, Ellerth focused all her attention in the Court of Appeals on proving her claim fit within that category. Given our explanation that the labels quid pro quo and hostile work environment are not controlling for purposes of establishing employer liability, see supra, at 8, Ellerth should have an adequate opportunity to prove she has a claim for which Burlington is liable.

Although Ellerth has not alleged she suffered a tangible employment action at the hands of Slowik, which would deprive Burlington of the availability of the affirmative defense, this is not dispositive. In light of our decision, Burlington is still subject to vicarious liability for Slowik's activity, but Burlington should have an opportunity to assert and prove the affirmative defense to liability. See supra, at 20–21.

For these reasons, we will affirm the judgment of the Court of Appeals, reversing the grant of summary judgment against Ellerth. On remand, the District Court will have the opportunity to decide whether it would be appropriate to allow Ellerth to amend her pleading or supplement her discovery.

The judgment of the Court of Appeals is affirmed.

It is so ordered.

U.S. Supreme Court

No. 97-569

BURLINGTON INDUSTRIES, INC., PETITIONER
v. KIMBERLY B. ELLERTH

ON WRIT OF CERTIORARI TO THE UNITED STATES
COURT OF APPEALS FOR THE SEVENTH CIRCUIT

[June 26, 1998]

JUSTICE GINSBURG, concurring in the judgment.

I agree with the Court's ruling that "the labels quid pro quo and hostile work environment are not controlling for purposes of establishing employer liability." Ante, at 2021. I also subscribe to the Court's statement of the rule governing employer liability, ante, at 20, which is substantively identical to the rule the Court adopts in Faragher v. Boca Raton, post, p. ___ .

U.S. Supreme Court

No. 97-569

BURLINGTON INDUSTRIES, INC., PETITIONER
v. KIMBERLY B. ELLERTH

ON WRIT OF CERTIORARI TO THE UNITED STATES
COURT OF APPEALS FOR THE SEVENTH CIRCUIT

[June 26, 1998]

JUSTICE THOMAS, with whom JUSTICE SCALIA joins, dissenting.

The Court today manufactures a rule that employers are vicariously liable if supervisors create a sexually hostile work environment, subject to an affirmative defense that the Court barely attempts to define. This rule applies even if the employer has a policy against sexual harassment, the employee knows about that policy, and the employee never informs anyone in a position of authority about the supervisor's conduct. As a result, employer liability under Title VII is judged by different standards depending upon whether a sexually or racially hostile work environment is alleged. The standard of employer liability should be the same in both instances: An employer should be liable if, and only if, the plaintiff proves that the employer was negligent in permitting the supervisor's conduct to occur.

I

Years before sexual harassment was recognized as "discriminat[ion] . . . because of . . . sex," 42 U.S.C. § 2000e2(a)(1), the Courts of Appeals considered whether, and when, a racially hostile work environment could violate Title VII.[1]

In the landmark case Rogers v. EEOC, 454 F. 2d 234 (1971), cert. denied, 406 U.S. 957 (1972), the Court of Appeals for the Fifth Circuit held that the practice of racially segregating patients in a doctor's office could amount to discrimination in " 'the terms, conditions, or privileges' " of employment, thereby violating Title VII. Id., at 238 (quoting 42 U.S.C. § 2000e-2(a)(1)). The principal opinion in the case concluded that employment discrimination was not limited to the "isolated and distinguishable events" of "hiring, firing, and promoting." Id., at 238 (opinion of Goldberg, J.). Rather, Title VII could also be violated by a work environment "heavily polluted with discrimination," because of the deleterious effects of such an atmosphere on an employee's well-being. Ibid.

Accordingly, after Rogers, a plaintiff claiming employment discrimination based upon race could assert a claim for a racially hostile work environment, in addition to the classic claim of so-called "disparate treatment." A disparate treatment claim required a plaintiff to prove an adverse employment consequence and discriminatory intent by his employer. See 1 B. Lindemann & P. Grossman, Employment Discrimination Law 10–11 (3d ed. 1996). A hostile environment claim required the plaintiff to show that his work environment was so pervaded by racial harassment as to alter the terms and conditions of his employment. See, e.g., Snell v. Suffolk Cty., 782 F. 2d 1094, 1103 (CA2 1986) ("To establish a hostile atmosphere, . . . plaintiffs must prove more than a few isolated incidents of racial enmity"); Johnson v. Bunny Bread Co., 646 F. 2d 1250, 1257 (CA8 1981) (no violation of Title VII from infrequent use of racial slurs). This is the same standard now used when determining whether sexual harassment renders a work environment hostile. See Harris v. Forklift Systems, Inc., 510 U.S. 17, 21 (1993) (actionable sexual harassment occurs when the workplace is "permeated with discriminatory intimidation, ridicule, and insult") (emphasis added) (internal quotation marks and citation omitted).

In race discrimination cases, employer liability has turned on whether the plaintiff has alleged an adverse employment consequence, such as firing or demotion, or a hostile work environment. If a supervisor takes an adverse employment action because of race, causing the employee a tangible job detriment, the employer is vicariously liable for resulting damages. See ante, at 15. This is because such actions are company acts that can be performed only by the exercise of specific authority granted by the employer, and thus the supervisor acts as the employer. If, on the other hand, the employee alleges a racially hostile work environment, the employer is liable only for negligence: that is, only if the employer knew, or in the exercise of reasonable care should have known, about the harassment and failed to take remedial action. See, e.g., Dennis v. Cty. of Fairfax, 55 F. 3d 151, 153 (CA4 1995); Davis v. Monsanto Chemical Co., 858 F. 2d 345, 349 (CA6 1988), cert. denied, 490 U.S. 1110 (1989). Liability has thus been imposed only if the employer is blameworthy in some way. See, e.g., Davis v. Monsanto Chemical Co., supra, at 349; Snell v. Suffolk Cty., supra, at 1104; DeGrace v. Rumsfeld, 614 F. 2d 796, 805 (CA1 1980).

This distinction applies with equal force in cases of sexual harassment.[2]

When a supervisor inflicts an adverse employment consequence upon an employee who has rebuffed his advances, the supervisor exercises the specific authority granted to him by his company. His acts, therefore, are the company's acts and are properly chargeable to it. See 123 F. 3d 490, 514 (1997) (Posner, C. J., dissenting); ante, at 17 ("Tangible employment actions fall within the special province of the supervisor. The supervisor has been empowered by the company as a distinct class of agent to make economic decisions affecting other employees under his or her control").

If a supervisor creates a hostile work environment, however, he does not act for the employer. As the Court concedes, a supervisor's creation of a hostile work environment is neither within the scope of his employment, nor part of his apparent authority. See ante, at 10–14. Indeed, a hostile work environment is antithetical to the interest of the employer. In such circumstances, an employer should be liable only if it has been negligent. That is, liability should attach only if the employer either knew, or in the exercise of reasonable care should have known, about the hostile work environment and failed to take remedial action.[3] Sexual harassment is simply not something that employers can wholly prevent without taking extraordinary measures—constant video and audio surveillance, for example—that would revolutionize the workplace in a manner incompatible with a free society. See 123 F. 3d 490, 513 (Posner, C.J., dissenting). Indeed, such measures could not even detect incidents of harassment such as the comments Slowick allegedly made to respondent in a hotel bar. The most that employers can be charged with, therefore, is a duty to act reasonably under the circumstances. As one court recognized in addressing an early racial harassment claim:

> "It may not always be within an employer's power to guarantee an environment free from all bigotry. . . . [H]e can let it be known, however, that racial harassment will not be tolerated, and he can take all reasonable measures to enforce this policy. . . . But once an employer has in good faith taken those measures which are both feasible and reasonable under the circumstances to combat the offensive conduct we do not think he can be charged with discriminating on the basis of race."

De Grace v. Rumsfeld, 614 F. 2d 796, 805 (1980).

Under a negligence standard, Burlington cannot be held liable for Slowick's conduct. Although respondent alleged a hostile work environment, she never contended that Burlington had been negligent in permitting the harassment to occur, and there is no question that Burlington acted reasonably under the circumstances. The company had a policy against sexual harassment, and respondent admitted that she was aware of the policy but nonetheless failed to tell anyone with authority over Slowick about his behavior. See, ante, at 3. Burlington therefore cannot be charged with knowledge of Slowick's alleged harassment or with a failure to exercise reasonable care in not knowing about it.

II

Rejecting a negligence standard, the Court instead imposes a rule of vicarious employer liability, subject to a vague affirmative defense, for the acts of supervisors who wield no delegated authority in creating a hostile work environment. This rule is a whole-cloth creation that draws no support from the legal principles on which the Court claims it is based. Compounding its error, the Court fails to explain how employers can rely upon the affirmative defense, thus ensuring a continuing reign of confusion in this important area of the law.

In justifying its holding, the Court refers to our comment in Meritor Savings Bank, FSB v. Vinson, 477 U.S. 57 (1986), that the lower courts should look to "agency principles" for guidance in determining the scope of employer liability, id., at 72. The Court then interprets the term "agency principles" to mean the Restatement (Second) of Agency (1957). The Court finds two portions of the Restatement to be relevant: § 219(2)(b), which provides that a master is liable for his servant's torts if the master is reckless or negligent, and § 219(2)(d), which states that a master is liable for his servant's torts when the servant is "aided in accomplishing the tort by the existence of the agency relation." The Court appears to reason that a supervisor is "aided . . . by . . . the agency relation" in creating a hostile work environment because the supervisor's "power and authority invests his or her harassing conduct with a particular threatening character." Ante, at 18.

Section 219(2)(d) of the Restatement provides no basis whatsoever for imposing vicarious liability for a supervisor's creation of a hostile work environment. Contrary to the Court's suggestions, the principle embodied in § 219(2)(d) has nothing to do with a servant's "power and authority," nor with whether his actions appear "threatening." Rather, as demonstrated by the Restatement's illustrations, liability under § 219(2)(d) depends upon the plaintiff's belief that the agent acted in the ordinary course of business or within the scope of his apparent authority.[4]

In this day and age, no sexually harassed employee can reasonably believe that a harassing supervisor is conducting the official business of the company or acting on its behalf. Indeed, the Court admits as much in demonstrating why sexual harassment is not committed within the scope of a supervisor's employment and is not part of his apparent authority. See ante, at 10–14.

Thus although the Court implies that it has found guidance in both precedent and statute—see ante, at 9 ("The resulting federal rule, based on a body of case law developed over time, is statutory interpretation pursuant to congressional direction")—its holding is a product of willful policymaking, pure and simple. The only agency principle that justifies imposing employer liability in this context is the principle that a master will be liable for a servant's torts if the master was negligent or reckless in permitting them to occur; and as noted, under a negligence standard, Burlington cannot be held liable. See supra, at 5–6.

The Court's decision is also in considerable tension with our holding in Meritor that employers are not strictly liable for a supervisor's sexual harassment. See Meritor Savings Bank, FSB v. Vinson, supra, at 72. Although the Court recognizes an affirmative defense—based solely on its divination of Title VII's gestalt, see ante, at 19—it provides shockingly little guidance about how employers can actually avoid vicarious liability. Instead, it issues only Delphic pronouncements and leaves the dirty work to the lower courts:

> "While proof that an employer had promulgated an anti-harassment policy with complaint procedure is not necessary in every instance as a matter of law, the need for a stated policy suitable to the employment circumstances may appropriately be addressed in any case when litigating the first element of the defense. And while proof that an employee failed to fulfill the corresponding obligation of reason-

able care to avoid harm is not limited to showing any unreasonable failure to use any complaint procedure provided by the employer, a demonstration of such failure will normally suffice to satisfy the employer's burden under the second element of the defense."

Ante, at 20.

What these statements mean for district courts ruling on motions for summary judgment—the critical question for employers now subject to the vicarious liability rule—remains a mystery. Moreover, employers will be liable notwithstanding the affirmative defense, even though they acted reasonably, so long as the plaintiff in question fulfilled her duty of reasonable care to avoid harm. See ibid. In practice, therefore, employer liability very well may be the rule. But as the Court acknowledges, this is the one result that it is clear Congress did not intend. See ante, at 18; Meritor Savings Bank, FSB v. Vinson, 477 U.S., at 72.

The Court's holding does guarantee one result: There will be more and more litigation to clarify applicable legal rules in an area in which both practitioners and the courts have long been begging for guidance. It thus truly boggles the mind that the Court can claim that its holding will effect "Congress' intention to promote conciliation rather than litigation in the Title VII context." Ante, at 19. All in all, today's decision is an ironic result for a case that generated eight separate opinions in the Court of Appeals on a fundamental question, and in which we granted certiorari "to assist in defining the relevant standards of employer liability." Ante, at 5.

Popular misconceptions notwithstanding, sexual harassment is not a freestanding federal tort, but a form of employment discrimination. As such, it should be treated no differently (and certainly no better) than the other forms of harassment that are illegal under Title VII. I would restore parallel treatment of employer liability for racial and sexual harassment and hold an employer liable for a hostile work environment only if the employer is truly at fault. I therefore respectfully dissent.

Notes

1. This sequence of events is not surprising, given that the primary goal of the Civil Rights Act of 1964 was to eradicate race discrimination and that the statute's ban on sex discrimination was added as an eleventh-hour amendment in an effort to kill the bill. See Barnes v. Costle, 561 F. 2d 983, 987 (CADC 1977).

2. The Courts of Appeals relied on racial harassment cases when analyzing early claims of discrimination based upon a supervisor's sexual harassment. For example, when the Court of Appeals for the District Columbia Circuit held that a work environment poisoned by a supervisor's "sexually stereotyped insults and demeaning propositions" could itself violate Title VII, its principal authority was Judge Goldberg's opinion in Rogers. See Bundy v. Jackson, 641 F. 2d 934, 944 (CADC 1981); see also Henson v. Dundee, 682 F. 2d 897, 901 (CA11 1982). So too, this Court relied on Rogers when in Meritor Savings Bank, FSB v. Vinson, 477 U.S. 57

(1986), it recognized a cause of action under Title VII for sexual harassment. See id., at 65–66.

3. I agree with the Court that the doctrine of quid pro quo sexual harassment is irrelevant to the issue of an employer's vicarious liability. I do not, however, agree that the distinction between hostile work environment and quid pro quo sexual harassment is relevant "when there is a threshold question whether a plaintiff can prove discrimination in violation of Title VII." Ante, at 8. A supervisor's threat to take adverse action against an employee who refuses his sexual demands, if never carried out, may create a hostile work environment, but that is all. Cases involving such threats, without more, should therefore be analyzed as hostile work environment cases only. If, on the other hand, the supervisor carries out his threat and causes the plaintiff a job detriment, the plaintiff may have a disparate treatment claim under Title VII. See E. Scalia, The Strange Career of Quid Pro Quo Sexual Harassment, 21 Harv. J. L. & Pub. Policy 307, 309–314 (1998).

4. See Restatement § 219, Comment e; § 261, Comment a (principal liable for an agent's fraud if "the agent's position facilitates the consummation of the fraud, in that from the point of view of the third person the transaction seems regular on its face and the agent appears to be acting in the ordinary course of business confided to him"); § 247, Illustrations (newspaper liable for a defamatory editorial published by editor for his own purposes).

- [U.S.]
- [U.S.]
- [U.S.]

APPENDIX C

Kolstad v. American Dental Association

CERTIORARI TO THE UNITED STATES COURT OF APPEALS FOR THE DISTRICT OF COLUMBIA CIRCUIT

No. 98-208.
Argued March 1, 1999–
Decided June 22, 1999

Petitioner sued respondent under Title VII of the Civil Rights Act of 1964 (Title VII), asserting that respondent's decision to promote Tom Spangler over her was a proscribed act of gender discrimination. Petitioner alleged, and introduced testimony to prove, that, among other things, the entire selection process was a sham, the stated reasons of respondent's executive director for selecting Spangler were pretext, and Spangler had been chosen before the formal selection process began. The District Court denied petitioner's request for a jury instruction on punitive damages, which are authorized by the Civil Rights Act of 1991 (1991 Act) for Title VII cases in which the employee "demonstrates" that the employer has engaged in intentional discrimination and has done so "with malice or with reckless indifference to [the employee's] federally protected rights." 42 U.S.C. § 1981a(b)(1). In affirming that denial, the en banc Court of Appeals concluded that, before the jury can be instructed on punitive damages, the evidence must demonstrate that the defendant has engaged in some "egregious" misconduct, and that petitioner had failed to make the requisite showing in this case.

Held:

1. An employer's conduct need not be independently "egregious" to satisfy § 1981a's requirements for a punitive damages award, although evidence of egregious behavior may provide a valuable means by which an employee can show the "malice" or "reckless indifference" needed to qualify for such

an award. The 1991 Act provided for compensatory and punitive damages in addition to the backpay and other equitable relief to which prevailing Title VII plaintiffs had previously been limited. Section 1981a's two-tiered structure—it limits compensatory and punitive awards to cases of "intentional discrimination," § 1981a(a)(1), and further qualifies the availability of punitive awards to instances of "malice" or "reckless indifference"—suggests a congressional intent to impose two standards of liability, one for establishing a right to compensatory damages and another, higher standard that a plaintiff must satisfy to qualify for a punitive award. The terms "malice" and "reckless indifference" ultimately focus on the actor's state of mind, however, and § 1981a does not require a showing of egregious or outrageous discrimination independent of the employer's state of mind. Nor does the statute's structure imply an independent role for "egregiousness" in the face of congressional silence. On the contrary, the view that § 1981a provides for punitive awards based solely on an employer's state of mind is consistent with the 1991 Act's distinction between equitable and compensatory relief. Intent determines which remedies are open to a plaintiff here as well. This focus on the employer's state of mind does give effect to the statute's two-tiered structure. The terms "malice" and "reckless indifference" pertain not to the employer's awareness that it is engaging in discrimination, but to its knowledge that it may be acting in violation of federal law, see, *e.g., Smith v. Wade,* 461 U.S. 30, 37, n. 6, 41, 50. There will be circumstances where intentional discrimination does not give rise to punitive damages liability under this standard, as where the employer is unaware of the relevant federal prohibition or discriminates with the distinct belief that its discrimination is lawful, where the underlying theory of discrimination is novel or otherwise poorly recognized, or where the employer reasonably believes that its discrimination satisfies a bona fide occupational qualification defense or other statutory exception to liability. See *Hazen Paper Co. v. Biggins,* 507 U.S. 604, 616, 617. Although there is some support for respondent's assertion that the common law punitive awards tradition includes an "egregious misconduct" requirement, eligibility for such awards most often is characterized in terms of a defendant's evil motive or intent. Egregious or outrageous acts may serve as evidence supporting an inference of such evil motive, but § 1981a does not limit plaintiffs to this form of evidence or require a showing of egregious or outrageous discrimination independent of the employer's state of mind. Pp. 5–11.

2. The inquiry does not end with a showing of the requisite mental state by certain employees, however. Petitioner must impute liability for punitive damages to respondent. Common law limitations on a principal's vicarious liability for its agents' acts apply in the Title VII context. See, *e.g., Burlington Industries, Inc. v. Ellerth,* 524 U.S. 742, 754. The Court's discussion of this question is informed by the general common law of agency, as codified in the Restatement (Second) of Agency, see, *e.g., id.,* at 755, which, among

other things, authorizes punitive damages "against a . . . principal because of an [agent's] act . . . if . . . the agent was employed in a managerial capacity and was acting in the scope of employment," § 217 C(c), and declares that even intentional, specifically forbidden torts are within such scope if the conduct is "the kind [the employee] is employed to perform," "occurs substantially within the authorized time and space limits," and "is actuated, at least in part, by a purpose to serve the" employer, §§ 228(1), 230, Comment *b*. Under these rules, even an employer who made every good faith effort to comply with Title VII would be held liable for the discriminatory acts of agents acting in a "managerial capacity." Holding such an employer liable, however, is in some tension with the principle that it is "improper . . . to award punitive damages against one who himself is personally innocent and therefore liable only vicariously," Restatement (Second) of Torts, § 909, Comment *b*. Applying the Restatement's "scope of employment" rule in this context, moreover, would reduce the incentive for employers to implement antidiscrimination programs and would, in fact, likely exacerbate employers' concerns that 42 U.S.C. § 1981a's "malice" and "reckless indifference" standard penalizes those employers who educate themselves and their employees on Title VII's prohibitions. Dissuading employers from implementing programs or policies to prevent workplace discrimination is directly contrary to Title VII's prophylactic purposes. See, *e.g., Burlington Industries, Inc.*, 524 U.S., at 764. Thus, the Court is compelled to modify the Restatement rules to avoid undermining Title VII's objectives. See, *e.g., ibid*. The Court therefore agrees that, in the punitive damages context, an employer may not be vicariously liable for the discriminatory employment decisions of managerial agents where these decisions are contrary to the employer's good faith efforts to comply with Title VII. Pp. 11–18.

3. The question whether petitioner can identify facts sufficient to support an inference that the requisite mental state can be imputed to respondent is left for remand. The parties have not yet had an opportunity to marshal the record evidence in support of their views on the application of agency principles in this case, and the en banc Court of Appeals had no reason to resolve the issue because it concluded that petitioner had failed to demonstrate the requisite "egregious" misconduct. Pp. 18–19.

139 F. 3d 958, vacated and remanded.

O'Connor, J., delivered the opinion of the Court, Part I of which was unanimous, Part II-A of which was joined by *Stevens, Scalia, Kennedy, Souter, Ginsburg*, and *Breyer* JJ., and Part II-B of which was joined by *Rehnquist*, C. J., and *Scalia, Kennedy*, and *Thomas*, JJ. *Rehnquist*, C. J., filed an opinion concurring in part and dissenting in part, in which *Thomas*, J., joined. *Stevens*, J., filed an opinion concurring in part and dissenting in part, in which *Souter, Ginsburg*, and *Breyer*, JJ., joined.

CAROLE KOLSTAD, PETITIONER v.
AMERICAN DENTAL ASSOCIATION

ON WRIT OF CERTIORARI TO THE UNITED STATES COURT
OF APPEALS FOR THE DISTRICT OF COLUMBIA CIRCUIT

[June 22, 1999]

Justice O'Connor delivered the opinion of the Court.

Under the terms of the Civil Rights Act of 1991 (1991 Act), 105 Stat. 1071, punitive damages are available in claims under Title VII of the Civil Rights Act of 1964 (Title VII), 78 Stat. 253, as amended, 42 U.S.C. § 2000e *et seq.* (1994 ed. and Supp. III), and the Americans with Disabilities Act of 1990 (ADA), 104 Stat. 328, 42 U.S.C. § 12101 *et seq.* Punitive damages are limited, however, to cases in which the employer has engaged in intentional discrimination and has done so "with malice or with reckless indifference to the federally protected rights of an aggrieved individual." Rev. Stat. § 1977, as amended, 42 U.S.C. § 1981a(b)(1). We here consider the circumstances under which punitive damages may be awarded in an action under Title VII.

I

A

In September 1992, Jack O'Donnell announced that he would be retiring as the Director of Legislation and Legislative Policy and Director of the Council on Government Affairs and Federal Dental Services for respondent, American Dental Association (respondent or Association). Petitioner, Carole Kolstad, was employed with O'Donnell in respondent's Washington, D.C., office, where she was serving as respondent's Director of Federal Agency Relations. When she learned of O'Donnell's retirement, she expressed an interest in filling his position. Also interested in replacing O'Donnell was Tom Spangler, another employee in respondent's Washington office. At this time, Spangler was serving as the Association's Legislative Counsel, a position that involved him in respondent's legislative lobbying efforts. Both petitioner and Spangler had worked directly with O'Donnell, and both had received "distinguished" performance ratings by the acting head of the Washington office, Leonard Wheat.

Both petitioner and Spangler formally applied for O'Donnell's position, and Wheat requested that Dr. William Allen, then serving as respondent's Executive Director in the Association's Chicago office, make the ultimate promotion decision. After interviewing both petitioner and Spangler, Wheat recommended that Allen select Spangler for O'Donnell's post. Allen notified petitioner in December 1992 that he had, in fact, selected Spangler to serve as O'Donnell's replacement. Petitioner's challenge to this employment decision forms the basis of the instant action.

B

After first exhausting her avenues for relief before the Equal Employment Opportunity Commission, petitioner filed suit against the Association in Federal District Court, alleging that respondent's decision to promote Spangler was an act of employment discrimination proscribed under Title VII. In petitioner's view, the entire selection process was a sham. Tr. 8 (Oct. 26, 1995) (closing argument for plaintiff's counsel). Counsel for petitioner urged the jury to conclude that Allen's stated reasons for selecting Spangler were pretext for gender discrimination, *id.*, at 19, 24, and that Spangler had been chosen for the position before the formal selection process began, *id.*, at 19. Among the evidence offered in support of this view, there was testimony to the effect that Allen modified the description of O'Donnell's post to track aspects of the job description used to hire Spangler. See *id.*, at 132–136 (Oct. 19, 1995) (testimony of Cindy Simms); *id.*, at 48–51 (Oct. 20, 1995) (testimony of Leonard Wheat). In petitioner's view, this "preselection" procedure suggested an intent by the Association to discriminate on the basis of sex. *Id.*, at 24. Petitioner also introduced testimony at trial that Wheat told sexually offensive jokes and that he had referred to certain prominent professional women in derogatory terms. See *id.*, at 120–124 (Oct. 18, 1995) (testimony of Carole Kolstad). Moreover, Wheat allegedly refused to meet with petitioner for several weeks regarding her interest in O'Donnell's position. See *id.*, at 112–113. Petitioner testified, in fact, that she had historically experienced difficulty gaining access to meet with Wheat. See *id.*, at 114–115. Allen, for his part, testified that he conducted informal meetings regarding O'Donnell's position with both petitioner and Spangler, see *id.*, at 148 (Oct. 23, 1995), although petitioner stated that Allen did not discuss the position with her, see *id.*, at 127–128 (Oct. 18, 1995).

The District Court denied petitioner's request for a jury instruction on punitive damages. The jury concluded that respondent had discriminated against petitioner on the basis of sex and awarded her backpay totaling $52,718. App. 109–110. Although the District Court subsequently denied respondent's motion for judgment as a matter of law on the issue of liability, the court made clear that it had not been persuaded that respondent had selected Spangler over petitioner on the basis of sex, and the court denied petitioner's requests for reinstatement and for attorney's fees. 912 F. Supp. 13, 15 (DC 1996).

Petitioner appealed from the District Court's decisions denying her requested jury instruction on punitive damages and her request for reinstatement and attorney's fees. Respondent cross-appealed from the denial of its motion for judgment as a matter of law. In a split decision, a panel of the Court of Appeals for the District of Columbia Circuit reversed the District Court's decision denying petitioner's request for an instruction on punitive damages. 108 F. 3d 1431, 1435 (1997). In so doing, the court rejected respondent's claim that punitive damages are available under Title VII only in " 'extraordinarily egregious cases.'" *Id.*, at 1437. The panel reasoned that, "because 'the state of mind necessary to trigger liability for the wrong is at least as culpable as that required to make punitive damages applicable,'" *id.*, at 1438 (quoting *Rowlett* v. *Anheuser-Busch, Inc.*, 832 F. 2d 194, 205 (CA1 1987)), the fact that the jury could reasonably have found intentional discrimination meant that the jury should have been permitted to consider punitive damages. The court noted, however, that not all cases involving intentional discrimination would support a punitive damages award. 108 F. 3d, at 1438. Such an award might be improper, the panel reasoned, in instances where the employer justifiably believes that intentional discrimination is permitted or where an employee engages in discrimination outside the scope of that employee's authority. *Id.*, at 1438–1439. Here, the court concluded, respondent "neither attempted to justify the use of sex in its promotion decision nor disavowed the actions of its agents." *Id.*, at 1439.

The Court of Appeals subsequently agreed to rehear the case en banc, limited to the punitive damages question. In a divided opinion, the court affirmed the decision of the District Court. 139 F. 3d 958 (1998). The en banc majority concluded that, "before the question of punitive damages can go to the jury, the evidence of the defendant's culpability must exceed what is needed to show intentional discrimination." *Id.*, at 961. Based on the 1991 Act's structure and legislative history, the court determined, specifically, that a defendant must be shown to have engaged in some "egregious" misconduct before the jury is permitted to consider a request for punitive damages. *Id.*, at 965. Although the court declined to set out the "egregiousness" requirement in any detail, it concluded that petitioner failed to make the requisite showing in the instant case. Judge Randolph concurred, relying chiefly on § 1981a's structure as evidence of a congressional intent to "limi[t] punitive damages to exceptional cases." *Id.*, at 970. Judge Tatel wrote in dissent for five judges, who agreed generally with the panel majority.

We granted certiorari, 525 U.S. ___ (1998), to resolve a conflict among the Federal Courts of Appeals concerning the circumstances under which a jury may consider a request for punitive damages under § 1981a(b)(1). Compare 139 F. 3d 958 (CADC 1998) (case below), with *Luciano* v. *Olsten Corp.*, 110 F. 3d 210, 219–220 (CA2 1997) (rejecting contention that punitive damages require showing of "extraordinarily egregious" conduct).

II

A

Prior to 1991, only equitable relief, primarily backpay, was available to prevailing Title VII plaintiffs; the statute provided no authority for an award of punitive or compensatory damages. See *Landgraf* v. *USI Film Products*, 511 U.S. 244, 252–253 (1994). With the passage of the 1991 Act, Congress provided for additional remedies, including punitive damages, for certain classes of Title VII and ADA violations.

The 1991 Act limits compensatory and punitive damages awards, however, to cases of "intentional discrimination"—that is, cases that do not rely on the "disparate impact" theory of discrimination. 42 U.S.C. § 1981a(a)(1). Section 1981a(b)(1) further qualifies the availability of punitive awards:

> "A complaining party may recover punitive damages under this section against a respondent (other than a government, government agency or political subdivision) if the complaining party demonstrates that the respondent engaged in a discriminatory practice or discriminatory practices *with malice or with reckless indifference to the federally protected rights of an aggrieved individual.*" (Emphasis added.)

The very structure of § 1981a suggests a congressional intent to authorize punitive awards in only a subset of cases involving intentional discrimination. Section 1981a(a)(1) limits compensatory and punitive awards to instances of intentional discrimination, while § 1981a(b)(1) requires plaintiffs to make an additional "demonstrat[ion]" of their eligibility for punitive damages. Congress plainly sought to impose two standards of liability—one for establishing a right to compensatory damages and another, higher standard that a plaintiff must satisfy to qualify for a punitive award.

The Court of Appeals sought to give life to this two-tiered structure by limiting punitive awards to cases involving intentional discrimination of an "egregious" nature. We credit the en banc majority's effort to effectuate congressional intent, but, in the end, we reject its conclusion that eligibility for punitive damages can only be described in terms of an employer's "egregious" misconduct. The terms "malice" and "reckless" ultimately focus on the actor's state of mind. See, *e.g.*, Black's Law Dictionary 956–957, 1270 (6th ed. 1990); see also W. Keeton, D. Dobbs, R. Keeton, & D. Owen, Prosser and Keeton, Law of Torts 212–214 (5th ed. 1984) (defining "willful," "wanton," and "reckless"). While egregious misconduct is evidence of the requisite mental state, see *infra,* at 10–11; Keeton, *supra,* at 213–214, § 1981a does not limit plaintiffs to this form of evidence, and the section does not require a showing of egregious or outrageous discrimination independent of the employer's state of mind. Nor does the statute's structure imply an independent role for "egregiousness" in the face of congressional silence. On the contrary, the view that § 1981a provides for punitive awards based solely on an employer's state of mind is consistent with the 1991 Act's distinction between equitable and compensatory relief. Intent determines which remedies are open to a plaintiff here as well; compensatory awards are available only where the employer has engaged in "*intentional* discrimination." § 1981a(a)(1) (emphasis added).

Moreover, § 1981a's focus on the employer's state of mind gives some effect to Congress' apparent intent to narrow the class of cases for which punitive awards are available to a subset of those involving intentional discrimination. The employer must act with "malice or with reckless indifference *to [the plaintiff's] federally protected rights.*" § 1981a(b)(1) (emphasis added). The terms "malice" or "reckless indifference" pertain to the employer's knowledge that it may be acting in violation of federal law, not its awareness that it is engaging in discrimination.

We gain an understanding of the meaning of the terms "malice" and "reckless indifference," as used in § 1981a, from this Court's decision in *Smith* v. *Wade,* 461 U.S. 30 (1983). The parties, as well as both the en banc majority and dissent, recognize that Congress looked to the Court's decision in *Smith* in adopting this language in § 1981a. See Tr. of Oral Arg. 28–29; Brief for Petitioner 24; 139 F. 3d, at 964–965; *id.,* at 971 (Tatel, J., dissenting). Employing language similar to what later appeared in § 1981a, the Court concluded in *Smith* that "a jury may be permitted to assess punitive damages in an action under § 1983 when the defendant's conduct is shown to be motivated by evil motive or intent, or when it involves reckless or callous indifference to the federally protected rights of others." 461 U.S., at 56. While the *Smith* Court determined that it was unnecessary to show actual malice to qualify for a punitive award, *id.,* at 45–48, its intent standard, at a minimum, required recklessness in its subjective form. The Court referred to a "subjective consciousness" of a risk of injury or illegality and a "'criminal indifference to civil obligations.'" *Id.,* at 37, n. 6, 41 (quoting *Philadelphia, W. & B. R. Co.* v. *Quigley,* 21 How. 202, 214 (1859)); see also *Farmer* v. *Brennan,* 511 U.S. 825, 837 (1994) (explaining that criminal law employs subjective form of recklessness, requiring a finding that the defendant "disregards a risk of harm of which he is aware"); see generally 1 T. Sedgwick, Measure of Damages §§ 366, 368, pp. 528, 529 (8th ed. 1891) (describing "wantonness" in punitive damages context in terms of "criminal indifference" and "gross negligence" in terms of a "conscious indifference to consequences"). The Court thus compared the recklessness standard to the requirement that defendants act with "'knowledge of falsity or reckless disregard for the truth'" before punitive awards are available in defamation actions, *Smith, supra,* at 50 (quoting *Gertz* v. *Robert Welch, Inc.,* 418 U.S. 323, 349 (1974)), a subjective standard, *Harte-Hanks Communications, Inc.* v. *Connaughton,* 491 U.S. 657, 688 (1989). Applying this standard in the context of § 1981a, an employer must at least discriminate in the face of a perceived risk that its actions will violate federal law to be liable in punitive damages.

There will be circumstances where intentional discrimination does not give rise to punitive damages liability under this standard. In some instances, the employer may simply be unaware of the relevant federal prohibition. There will be cases, moreover, in which the employer discriminates with the distinct belief that its discrimination is lawful. The underlying theory of discrimination may be novel or otherwise poorly recognized, or an employer may reasonably believe that its discrimination satisfies a bona fide occupational qualification defense or other statutory exception to liability. See, *e.g.*, 42 U.S.C. § 2000e-2(e)(1) (setting out Title VII defense "where religion, sex, or national origin is a bona fide occupational qualification"); see also § 12113 (setting out defenses under ADA). In *Hazen Paper Co.* v. *Biggins*, 507 U.S. 604, 616 (1993), we thus observed that, in light of statutory defenses and other exceptions permitting age-based decisionmaking, an employer may knowingly rely on age to make employment decisions without recklessly violating the Age Discrimination in Employment Act of 1967 (ADEA). Accordingly, we determined that limiting liquidated damages under the ADEA to cases where the employer "knew or showed reckless disregard for the matter of whether its conduct was prohibited by the statute," without an additional showing of outrageous conduct, was sufficient to give effect to the ADEA's two-tiered liability scheme. *Id.*, at 616, 617.

At oral argument, respondent urged that the common law tradition surrounding punitive awards includes an "egregious misconduct" requirement. See, *e.g.*, Tr. of Oral Arg. 26–28; see also Brief for Chamber of Commerce of United States as *Amicus Curiae* 8–22 (advancing this argument). We assume that Congress, in legislating on punitive awards, imported common law principles governing this form of relief. See, *e.g.*, *Molzof* v. *United States*, 502 U.S. 301, 307 (1992). Moreover, some courts and commentators have described punitive awards as requiring both a specified state of mind and egregious or aggravated misconduct. See, *e.g.*, 1 D. Dobbs, Law of Remedies 468 (2d ed. 1993) ("Punitive damages are awarded when the defendant is guilty of both a bad state of mind and highly serious misconduct").

Most often, however, eligibility for punitive awards is characterized in terms of a defendant's motive or intent. See, *e.g.*, 1 Sedgwick, *supra,* at 526, 528; C. McCormick, Law of Damages 280 (1935). Indeed, "[t]he justification of exemplary damages lies in the evil intent of the defendant." 1 Sedgwick, *supra,* at 526; see also 2 J. Sutherland, Law of Damages § 390, p. 1079 (3d ed. 1903) (discussing punitive damages under rubric of "[c]ompensation for wrongs done with bad motive"). Accordingly, "a positive element of conscious wrongdoing is always required." McCormick, *supra,* at 280.

Egregious misconduct is often associated with the award of punitive damages, but the reprehensible character of the conduct is not generally considered apart from the requisite state of mind. Conduct warranting punitive awards has been characterized as "egregious," for example, *because* of the defendant's mental state. See Restatement (Second) of Torts § 908(2) (1979) ("Punitive damages may be awarded for conduct that is outrageous, because of the defendant's evil motive or his reckless indifference to the rights of others"). Respondent, in fact, appears to endorse this characterization. See, *e.g.*,

Brief for Respondent 19 ("Malicious and reckless conduct [is] by definition egregious"); see also *id.*, at 28–29. That conduct committed with the specified mental state may be characterized as egregious, however, is not to say that employers must engage in conduct with some independent, "egregious" quality before being subject to a punitive award.

To be sure, egregious or outrageous acts may serve as evidence supporting an inference of the requisite "evil motive." "The allowance of exemplary damages depends upon the bad motive of the wrong-doer *as exhibited by his acts.*" 1 Sedgwick, *supra,* at 529 (emphasis added); see also 2 Sutherland, *supra,* § 394, at 1101 ("The spirit which actuated the wrong-doer may doubtless be inferred from the circumstances surrounding the parties and the transaction"); see, *e.g., Chizmar v. Mackie,* 896 P. 2d 196, 209 (Alaska 1995) ("[W]here there is no evidence that gives rise to an inference of actual malice or conduct sufficiently outrageous to be deemed equivalent to actual malice, the trial court need not, and indeed should not, submit the issue of punitive damages to the jury" (internal quotation marks omitted)); *Horton v. Union Light, Heat & Power Co.,* 690 S. W. 2d 382, 389 (Ky. 1985) (observing that "malice . . . may be implied from outrageous conduct"). Likewise, under § 1981a(b)(1), pointing to evidence of an employer's egregious behavior would provide one means of satisfying the plaintiff's burden to "demonstrat[e]" that the employer acted with the requisite "malice or . . . reckless indifference." See 42 U.S.C. § 1981a(b)(1); see, *e.g.,* 3 BNA EEOC Compliance Manual N:6085–N6084 (1992) (Enforcement Guidance: Compensatory and Punitive Damages Available Under § 102 of the Civil Rights Act of 1991) (listing "[t]he degree of egregiousness and nature of the respondent's conduct" among evidence tending to show malice or reckless disregard). Again, however, respondent has not shown that the terms "reckless indifference" and "malice," in the punitive damages context, have taken on a consistent definition including an independent, "egregiousness" requirement. Cf. *Morissette v. United States,* 342 U.S. 246, 263 (1952) ("[W]here Congress borrows terms of art in which are accumulated the legal tradition and meaning of centuries of practice, it presumably knows and adopts the cluster of ideas that were attached to each borrowed word in the body of learning from which it was taken and the meaning its use will convey to the judicial mind unless otherwise instructed").

B

The inquiry does not end with a showing of the requisite "malice or . . . reckless indifference" on the part of certain individuals, however. 42 U.S.C. § 1981a(b)(1). The plaintiff must impute liability for punitive damages to respondent. The en banc dissent recognized that agency principles place limits on vicarious liability for punitive damages. 139 F. 3d, at 974 (Tatel, J., dissenting). Likewise, the Solicitor General as *amicus* acknowledged during argument that common law limitations on a principal's liability in punitive awards for the acts of its agents apply in the Title VII context. Tr. of Oral Arg. 23.

Justice Stevens urges that we should not consider these limitations here. See *post,* at 6–8 (opinion concurring in part and dissenting in part). While we decline to engage in any definitive application of the agency standards to the facts of this case, see *infra,* at 18, it is important that we address the proper legal standards for imputing liability to an employer in the punitive damages context. This issue is intimately bound up with the preceding discussion on the evidentiary showing necessary to qualify for a punitive award, and it is easily subsumed within the question on which we granted certiorari—namely, "[i]n what circumstances may punitive damages be awarded under Title VII of the 1964 Civil Rights Act, as amended, for unlawful intentional discrimination?" Pet. for Cert. i; see also this Court's Rule 14.1(a). "On a number of occasions, this Court has considered issues waived by the parties below and in the petition for certiorari because the issues were so integral to decision of the case that they could be considered 'fairly subsumed' by the actual questions presented." *Gilmer* v. *Interstate/Johnson Lane Corp.,* 500 U.S. 20, 37 (1991) (Stevens, J., dissenting) (citing cases). The Court has not always confined itself to the set of issues addressed by the parties. See, *e.g., Steel Co.* v. *Citizens for a Better Environment,* 523 U.S. 83, 93–102 and n. 1 (1998); *H. J. Inc.* v. *Northwestern Bell Telephone Co.,* 492 U.S. 229, 243–249 (1989); *Continental Ill. Nat. Bank & Trust Co. v. Chicago R. I. & P. R. Co.,* 294 U.S. 648, 667–675 (1935). Here, moreover, limitations on the extent to which principals may be liable in punitive damages for the torts of their agents was the subject of discussion by both the en banc dissent and majority, see 139 F. 3d, at 968; *id.,* at 974 (Tatel, J., dissenting), *amicus* briefing, see Brief for Chamber of Commerce of the United States 22–27, and substantial questioning at oral argument, see Tr. of Oral Arg. 11–17, 19–24, 49–50, 54–55. Nor did respondent discount the notion that agency principles may place limits on an employer's vicarious liability for punitive damages. See *post,* at 6. In fact, respondent advanced the general position "that the higher agency principles, under common law, would apply to punitive damages." Tr. of Oral Arg. 49. Accordingly, we conclude that these potential limitations on the extent of respondent's liability are properly considered in the instant case.

The common law has long recognized that agency principles limit vicarious liability for punitive awards. See, *e.g.,* G. Field, Law of Damages §§ 85–87 (1876); 1 Sedgwick, Damages § 378; McCormick, Damages § 80; 2 F. Mechem, Law of Agency §§ 2014–2015 (2d ed. 1914). This is a principle, moreover, that this Court historically has endorsed. See, *e.g., Lake Shore & Michigan Southern R. Co.* v. *Prentice,* 147 U.S. 101, 114–115 (1893); *The Amiable Nancy,* 3. Wheat. 546, 558–559 (1818). Courts of Appeals, too, have relied on these liability limits in interpreting 42 U.S.C. § 1981a. See, *e.g., Dudley* v. *Wal-Mart Stores, Inc.,* 166 F. 3d 1317, 1322–1323 (CA11 1999); *Harris, supra,* at 983–985. See also *Fitzgerald* v. *Mountain States*

Telephone & Telegraph Co., 68 F. 3d 1257, 1263–1264 (CA10 1995) (same in suit under 42 U.S.C. § 1981). But see *Deffenbaugh-Williams* v. *Wal-Mart Stores, Inc.*, 156 F. 3d 581, 592–594 (CA5 1998), rehearing en banc ordered, 169 F. 3d 215 (1999).

We have observed that, "[i]n express terms, Congress has directed federal courts to interpret Title VII based on agency principles." *Burlington Industries, Inc.* v. *Ellerth*, 524 U.S. 742, 754 (1998); see also *Meritor Savings Bank, FSB* v. *Vinson*, 477 U.S. 57, 72 (1986) (noting that, in interpreting Title VII, "Congress wanted courts to look to agency principles for guidance"). Observing the limits on liability that these principles impose is especially important when interpreting the 1991 Act. In promulgating the Act, Congress conspicuously left intact the "limits of employer liability" established in *Meritor. Faragher* v. *Boca Raton*, 524 U.S. 775, 804, n. 4 (1998); see also *Burlington Industries, Inc., supra,* at 763–764 ("[W]e are bound by our holding in *Meritor* that agency principles constrain the imposition of vicarious liability in cases of supervisory harassment").

Although jurisdictions disagree over whether and how to limit vicarious liability for punitive damages, see, *e.g.,* 2 J. Ghiardi & J. Kircher, Punitive Damages: Law and Practice § 24.01 (1998) (discussing disagreement); 22 Am. Jur. 2d, Damages § 788 (1988) (same), our interpretation of Title VII is informed by "the general common law of agency, rather than . . . the law of any particular State." *Burlington Industries, Inc., supra,* at 754 (internal quotation marks omitted). The common law as codified in the Restatement (Second) of Agency (1957), provides a useful starting point for defining this general common law. See *Burlington Industries, Inc., supra,* at 755 ("[T]he Restatement . . . is a useful beginning point for a discussion of general agency principles"); see also *Meritor, supra,* at 72. The Restatement of Agency places strict limits on the extent to which an agent's misconduct may be imputed to the principal for purposes of awarding punitive damages:

> "Punitive damages can properly be awarded against a master or other principal because of an act by an agent if, but only if:
>
> "(a) the principal authorized the doing and the manner of the act, or
>
> "(b) the agent was unfit and the principal was reckless in employing him, or
>
> "(c) the agent was employed in a managerial capacity and was acting in the scope of employment, or
>
> "(d) the principal or a managerial agent of the principal ratified or approved the act." Restatement (Second) of Agency, *supra,* § 217 C.

See also Restatement (Second) of Torts § 909 (same).

The Restatement, for example, provides that the principal may be liable for punitive damages if it authorizes or ratifies the agent's tortious act, or if it acts recklessly in employing the malfeasing agent. The Restatement also contemplates liability for punitive awards where an employee serving in a "managerial capacity" committed the wrong while "acting in the scope of employment." Restatement (Second) of Agency, *supra,* § 217 C; see also Restatement (Second) of Torts, *supra,* § 909 (same). "Unfortunately, no good definition of what constitutes a 'managerial capacity' has been found," 2 Ghiardi, *supra,* § 24.05, at 14, and determining whether an employee meets this description requires a fact-intensive inquiry, *id.,* § 24.05; 1 L. Schlueter & K. Redden, Punitive Damages, § 4.4(B)(2)(a), p. 182 (3d ed. 1995). "In making this determination, the court should review the type of authority that the employer has given to the employee, the amount of discretion that the employee has in what is done and how it is accomplished." *Id.,* § 4.4(B)(2)(a), at 181. Suffice it to say here that the examples provided in the Restatement of Torts suggest that an employee must be "important," but perhaps need not be the employer's "top management, officers, or directors," to be acting "in a managerial capacity." *Ibid.;* see also 2 Ghiardi, *supra,* § 24.05, at 14; Restatement (Second) of Torts, § 909, at 468, Comment *b* and Illus. 3.

Additional questions arise from the meaning of the "scope of employment" requirement. The Restatement of Agency provides that even intentional torts are within the scope of an agent's employment if the conduct is "the kind [the employee] is employed to perform," "occurs substantially within the authorized time and space limits," and "is actuated, at least in part, by a purpose to serve the" employer. Restatement (Second) of Agency, *supra,* § 228(1), at 504. According to the Restatement, so long as these rules are satisfied, an employee may be said to act within the scope of employment even if the employee engages in acts "specifically forbidden" by the employer and uses "forbidden means of accomplishing results." *Id.,* § 230, at 511, Comment *b;* see also *Burlington Industries, Inc., supra,* at 756; Keeton, Torts § 70. On this view, even an employer who makes every effort to comply with Title VII would be held liable for the discriminatory acts of agents acting in a "managerial capacity."

Holding employers liable for punitive damages when they engage in good faith efforts to comply with Title VII, however, is in some tension with the very principles underlying common law limitations on vicarious liability for punitive damages—that it is "improper ordinarily to award punitive damages against one who himself is personally innocent and therefore liable only vicariously." Restatement (Second) of Torts, *supra,* § 909, at 468, Comment *b.* Where an employer has undertaken such good faith efforts at Title VII compliance, it "demonstrat[es] that it never acted in reckless disregard of federally protected rights." 139 F. 3d, at 974 (Tatel, J., dissenting); see also *Harris,* 132 F. 3d, at 983, 984 (observing that, "[i]n some cases, the existence of a written policy instituted in good faith has operated as a total bar to employer liability for punitive damages" and concluding that "the institution of a written sexual harassment policy goes a long way towards dispelling any claim about the employer's 'reckless' or 'malicious' state of mind").

Applying the Restatement of Agency's "scope of employment" rule in the Title VII punitive damages context, moreover, would reduce the incentive for employers to implement antidiscrimination programs. In fact, such a rule would likely exacerbate concerns among employers that § 1981a's "malice" and "reckless indifference" standard penalizes those employers who educate themselves and their employees on Title VII's prohibitions. See Brief for Equal Employment Advisory Council as *Amicus Curiae* 12 ("[I]f an employer has made efforts to familiarize itself with Title VII's requirements, then any violation of those requirements by the employer can be inferred to have been committed 'with malice or with reckless indifference'"). Dissuading employers from implementing programs or policies to prevent discrimination in the workplace is directly contrary to the purposes underlying Title VII. The statute's "primary objective" is "a prophylactic one," *Albemarle Paper Co. v. Moody,* 422 U.S. 405, 417 (1975); it aims, chiefly, "not to provide redress but to avoid harm," *Faragher,* 524 U.S., at 806. With regard to sexual harassment, "[f]or example, Title VII is designed to encourage the creation of antiharassment policies and effective grievance mechanisms." *Burlington Industries, Inc.,* 524 U.S., at 764. The purposes underlying Title VII are similarly advanced where employers are encouraged to adopt antidiscrimination policies and to educate their personnel on Title VII's prohibitions.

In light of the perverse incentives that the Restatement's "scope of employment" rules create, we are compelled to modify these principles to avoid undermining the objectives underlying Title VII. See generally *ibid.* See also *Faragher, supra,* at 802, n. 3 (noting that Court must "adapt agency concepts to the practical objectives of Title VII"); *Meritor Savings Bank, FSB,* 477 U.S., at 72 ("[C]ommon-law principles may not be transferable in all their particulars to Title VII"). Recognizing Title VII as an effort to promote prevention as well as remediation, and observing the very principles underlying the Restatements' strict limits on vicarious liability for punitive damages, we agree that, in the punitive damages context, an employer may not be vicariously liable for the discriminatory employment decisions of managerial agents where these decisions are contrary to the employer's "good-faith efforts to comply with Title VII." 139 F. 3d, at 974 (Tatel, J., dissenting). As the dissent recognized, "[g]iving punitive damages protection to employers who make good-faith efforts to prevent discrimination in the workplace accomplishes" Title VII's objective of "motivat[ing] employers to detect and deter Title VII violations." *Ibid.*

We have concluded that an employer's conduct need not be independently "egregious" to satisfy § 1981a's requirements for a punitive damages award, although evidence of egregious misconduct may be used to meet the plaintiff's burden of proof. We leave for remand the question whether petitioner can identify facts sufficient to support an inference that the requisite mental state can be imputed to respondent. The parties have not yet had an opportunity to marshal the record evidence in support of their views on the application of agency principles in the instant case, and the en banc majority had no reason to resolve the issue because it concluded that petitioner had failed to demonstrate the requisite "egregious" misconduct. 139 F. 3d, at 968. Although trial testimony established that Allen made the ultimate decision to promote Spangler while serving as petitioner's interim executive director, respondent's highest position, Tr. 159 (Oct. 19, 1995), it remains to be seen whether petitioner can make a sufficient showing that Allen acted with malice or reckless indifference to petitioner's Title VII rights. Even if it could be established that Wheat effectively selected O'Donnell's replacement, moreover, several questions would remain, e.g., whether Wheat was serving in a "managerial capacity" and whether he behaved with malice or reckless indifference to petitioner's rights. It may also be necessary to determine whether the Association had been making good faith efforts to enforce an antidiscrimination policy. We leave these issues for resolution on remand.

For the foregoing reasons, the decision of the Court of Appeals is vacated, and the case is remanded for proceedings consistent with this opinion.

It is so ordered.

CAROLE KOLSTAD, PETITIONER v.
AMERICAN DENTAL ASSOCIATION

ON WRIT OF CERTIORARI TO THE UNITED STATES COURT
OF APPEALS FOR THE DISTRICT OF COLUMBIA CIRCUIT

[June 22, 1999]

Chief Justice Rehnquist, with whom *Justice Thomas* joins, concurring in part and dissenting in part.

For the reasons stated by Judge Randolph in his concurring opinion in the Court of Appeals, I would hold that Congress' two-tiered scheme of Title VII monetary liability implies that there is an egregiousness requirement that reserves punitive damages only for the worst cases of intentional discrimination. See 139 F. 3d 958, 970 (CADC 1998). Since the Court has determined otherwise, however, I join that portion of Part II-B of the Court's opinion holding that principles of agency law place a significant limitation, and in many foreseeable cases a complete bar, on employer liability for punitive damages.

CAROLE KOLSTAD, PETITIONER v.
AMERICAN DENTAL ASSOCIATION

ON WRIT OF CERTIORARI TO THE UNITED STATES COURT
OF APPEALS FOR THE DISTRICT OF COLUMBIA CIRCUIT

[June 22, 1999]

Justice Stevens, with whom *Justice Souter, Justice Ginsburg,* and *Justice Breyer* join, concurring in part and dissenting in part.

The Court properly rejects the Court of Appeals' holding that defendants in Title VII actions must engage in "egregious" misconduct before a jury may be permitted to consider a request for punitive damages. Accordingly, I join Parts I and II-A of its opinion. I write separately, however, because I strongly disagree with the Court's decision to volunteer commentary on an issue that the parties have not briefed and that the facts of this case do not present. I would simply remand for a trial on punitive damages.

I

In enacting the Civil Rights Act of 1991 (1991 Act), Congress established a three-tiered system of remedies for a broad range of discriminatory conduct, including violations of Title VII of the Civil Rights Act of 1964, 42 U.S.C. § 2000e *et seq.*, as well as some violations of the Americans with Disabilities Act of 1990 (ADA), 42 U.S.C. § 12101 *et seq.* (1994 ed. and Supp II). Equitable remedies are available for disparate impact violations; compensatory damages for intentional disparate treatment; and punitive damages for intentional discrimination "with malice or with reckless indifference to the federally protected rights of an aggrieved individual." § 1981a(b)(1).

The 1991 Act's punitive damages standard, as the Court recognizes, *ante*, at 7, is quite obviously drawn from our holding in *Smith* v. *Wade*, 461 U.S. 30 (1983). There, we held that punitive damages may be awarded under 42 U.S.C. § 1983 (1976 ed., Supp. V) "when the defendant's conduct is shown to be motivated by evil motive or intent, or when it involves reckless or callous indifference to the federally protected rights of others." 461 U.S., at 56.[1] The 1991 Act's standard is also the same intent-based standard used in the Age Discrimination in Employment Act of 1967 (ADEA), 29 U.S.C. § 621 *et seq.* (1994 ed. and Supp. II). The ADEA provides for an award of liquidated damages—damages that are "punitive in nature," *Trans World Airlines, Inc.* v. *Thurston*, 469 U.S. 111, 125 (1985)—when the employer "knew or showed reckless disregard for the matter of whether its conduct was prohibited by the statute." *Hazen Paper Co.* v. *Biggins,* 507 U.S. 604, 617 (1993); accord, *Thurston,* 469 U.S., at 126.

In *Smith,* we carefully noted that our punitive damages standard separated the "quite distinct concepts of *intent to cause* injury, on one hand, and *subjective consciousness* of risk of injury (or of unlawfulness) on the other," 461 U.S., at 38, n. 6, and held that punitive damages are permissible only when the latter component is satisfied by a deliberate or recklessly indifferent violation of federal law. In *Thurston,* we interpreted the ADEA's standard the same way and explained that the relevant mental distinction between intentional discrimination and "reckless disregard" for federally protected rights is essentially the same as the well-known difference between a "knowing" and a "willful" violation of a criminal law. See 469 U.S., at 126–127. While a criminal defendant, like an employer, need not have knowledge of the law to act "knowingly" or intentionally, he must know that his acts violate the law or must "careless[ly] disregard whether or not one has the right so to act" in order to act "willfully." *United States* v. *Murdock,* 290 U.S. 389, 395 (1933), quoted in *Thurston,* 469 U.S., at 127. We have interpreted the word "willfully" the same way in the civil context. See *McLaughlin* v. *Richland Shoe Co.,* 486 U.S. 128, 133 (1988) (holding that the "plain language" of the Fair Labor Standards Act's "willful" liquidated damages standard requires that "the employer either knew or showed reckless disregard for the matter of whether its conduct was prohibited by the statute," without regard to the outrageousness of the conduct at issue).

Construing § 1981a(b)(1) to impose a purely mental standard is perfectly consistent with the structure and purpose of the 1991 Act. As with the ADEA, the 1991 Act's "willful" or "reckless disregard" standard respects the Act's "two-tiered" damages scheme while deterring future intentionally unlawful discrimination. See *Hazen Paper,* 507 U.S., at 614–615. There are, for reasons the Court explains, see *ante,* at 8–9, numerous instances in which an employer might intentionally treat an individual differently because of her race, gender, religion, or disability without knowing that it is violating Title VII or the ADA. In order to recover compensatory damages under the 1991 Act, victims of unlawful disparate treatment must prove that the defendants' *conduct* was intentional, but they need not prove that the defendants either knew or should have known that they were *violating the law.* It is the additional element of willful or reckless disregard of the law that justifies a penalty of double damages in age discrimination cases and punitive damages in the broad range of cases covered by the 1991 Act.

It is of course true that as our society moves closer to the goal of eliminating intentional, invidious discrimination, the core mandates of Title VII and the ADA are becoming increasingly ingrained in employers' minds. As more employers come to appreciate the importance and the proportions of those statutes' mandates, the number of federal violations will continue to decrease accordingly. But at the same time, one could reasonably believe, as Congress did, that as our national resolve against employment discrimination hardens, deliberate violations of Title VII and the ADA become increasingly blameworthy and more properly the subject of "societal condemnation," *McKennon v. Nashville Banner Publishing Co.,* 513 U.S. 352, 357 (1995), in the form of punitive damages. Indeed, it would have been rather perverse for Congress to conclude that the increasing acceptance of antidiscrimination laws in the workplace somehow mitigates willful violations of those laws such that only those violations that are accompanied by particularly outlandish acts warrant special deterrence.

Given the clarity of our cases and the precision of Congress' words, the common-law tradition of punitive damages and any relationship it has to "egregious conduct" is quite irrelevant. It is enough to say that Congress provided in the 1991 Act its own punitive damages standard that focuses solely on willful mental state, and it did not suggest that there is any class of willful violations that are exempt from exposure to punitive damages. Nor did it indicate that there is a point on the spectrum of deliberate or recklessly indifferent conduct that qualifies as "egregious." Thus, while behavior that merits that opprobrious label may provide probative evidence of wrongful motive, it is not a necessary prerequisite to proving such a motive under the 1991 Act. To the extent that any treatise or federal, state, or "common-law" case might suggest otherwise, it is wrong.

There are other means of proving that an employer willfully violated the law. An employer, may, for example, express hostility toward employment discrimination laws or conceal evidence regarding its "true" selection procedures because it knows they violate federal law. Whatever the case, so long as a Title VII plaintiff proffers sufficient evidence from which a jury could conclude that an employer acted willfully, judges have no place making their own value judgments regarding whether the conduct was "egregious" or otherwise presents an inappropriate candidate for punitive damages; the issue must go to the jury.

If we accept the jury's appraisal of the evidence in this case and draw, as we must when reviewing the denial of a jury instruction, all reasonable inferences in petitioner's favor, there is ample evidence from which the jury could have concluded that respondent willfully violated Title VII. Petitioner emphasized, at trial and in her briefs to this Court, that respondent took "a tangible employment action" against her in the form of denying a promotion. Brief for Petitioner 47. Evidence indicated that petitioner was the more qualified of the two candidates for the job. Respondent's decisionmakers, who were senior executives of the Association, were known occasionally to tell sexually offensive jokes and referred to professional women in derogatory terms. The record further supports an inference that these executives not only deliberately refused to consider petitioner fairly and to promote her because she is a woman, but they manipulated the job requirements and conducted a "sham" selection procedure in an attempt to conceal their misconduct.

There is no claim that respondent's decisionmakers violated any company policy; that they were not acting within the scope of their employment; or that respondent has ever disavowed their conduct. Neither the respondent nor its two decisionmakers claimed at trial any ignorance of Title VII's requirements, nor did either offer any "good-faith" reason for believing that being a man was a legitimate requirement for the job. Rather, at trial respondent resorted to false, pretextual explanations for its refusal to promote petitioner.

The record, in sum, contains evidence from which a jury might find that respondent acted with reckless indifference to petitioner's federally protected rights. It follows, in my judgment, that the three-judge panel of the Court of Appeals correctly decided to remand the case to the district court for a trial on punitive damages. See 108 F. 3d 1431, 1440 (CADC 1997). To the extent that the Court's opinion fails to direct that disposition, I respectfully dissent.

II

In Part II-B of its opinion, the Court discusses the question "whether liability for punitive damages may be imputed to respondent" under "agency principles." *Ante,* at 12. That is a question that neither of the parties has ever addressed in this litigation and that respondent, at least, has expressly disavowed. When prodded at oral argument, counsel for respondent twice stood firm on this point. "[W]e all agree," he twice repeated, "that that precise issue is not before the Court." Tr. of Oral Arg. 49. Nor did any of the 11 judges in the Court of Appeals believe that it was applicable to the dispute at hand—presumably because promotion decisions are quintessential "company acts," see 139 F. 3d 958, 968 (CADC 1998), and because the two executives who made this promotion decision were the executive director of the Association and the acting head of its Washington office. *Id.,* at 974, 979 (Tatel, J., dissenting). See also 108 F. 3d, at 1434, 1439. Judge Tatel, who the Court implies raised the agency issue, in fact explicitly (and correctly) concluded that "[t]his case does not present these or analogous circumstances." 108 F. 3d, at 1439.

The absence of briefing or meaningful argument by the parties makes this Court's gratuitous decision to volunteer an opinion on this nonissue particularly ill advised. It is not this Court's practice to consider arguments—specifically, alternative defenses of the judgment under review—that were not presented in the brief in opposition to the petition for certiorari. See this Court's Rule 15.2. Indeed, on two occasions in this very Term, we refused to do so despite the fact that the issues were briefed and argued by the parties. See *South Central Bell Telephone Co. v. Alabama,* 526 U.S. ____ , ____ (1999) (slip op., at 10); *Roberts v. Galen of Virginia, Inc.,* 525 U.S. ____ ,

____ (1999) *(per curiam)* (slip op., at 4–5). If we declined to reach alternate defenses under those circumstances, surely we should do so here.

Nor is it accurate for the Court to imply that the Solicitor General as *amicus* advocates a course similar to that which the Court takes regarding the agency question. Cf. *ante,* at 12. The Solicitor General, like the parties, did not brief any agency issue. At oral argument, he correspondingly stated that the issue "is not really presented here." Tr. of Oral Arg. 19. He then responded to the Court's questions by stating that the Federal Government believes that whenever a tangible employment consequence is involved § 1981a incorporates the "managerial capacity" principles espoused by § 217C of the Restatement (Second) of Agency. See Tr. of Oral Arg. 23. But to the extent that the Court tinkers with the Restatement's standard, it is rejecting the Government's view of its own statute without giving it an opportunity to be heard on the issue.

Accordingly, while I agree with the Court's rejection of the en banc majority's holding on the only issue that it confronted, I respectfully dissent from the Court's failure to order a remand for trial on the punitive damages issue.

Note

1. Lest there be any doubt that Congress looked to *Smith* in crafting the statute, the Report of the House Judiciary Committee explains that the "standard for punitive damages is taken directly from civil rights case law," H. R. Rep. No. 102-40, pt. 2, p. 29 (1991) and proceeds to quote and cite with approval the very page in *Smith* that announced the punitive damages standard requiring "evil motive or intent, or . . . reckless or callous indifference to the federally protected rights of others," 461 U.S., at 56, quoted in H. R. Rep. No. 102-40, at 29. The Report of the House Education and Labor Committee echoed this sentiment. See H. R. Rep. No. 102-40, p. 74 (1991) (citing *Smith* with approval). Congress' substitution in the 1991 Act of the word "malice" for *Smith v. Wade*'s phrase "evil motive or intent" is inconsequential; in *Smith,* we noted that "malice . . . may be an appropriate" term to denote ill will or an intent to injure. See 461 U.S., at 37, n. 6.

- [U.S.]
- [U.S.]
- [U.S.]

APPENDIX D

Enforcement Guidance: Vicarious Employer Liability for Unlawful Harassment by Supervisors

1. *SUBJECT:* Enforcement Guidance: Vicarious Employer Liability for Unlawful Harassment by Supervisors

2. *PURPOSE:* This document provides guidance regarding employer liability for harassment by supervisors based on sex, race, color, religion, national origin, age, disability, or protected activity.

3. *EFFECTIVE DATE:* Upon receipt.

4. *EXPIRATION DATE:* As an exception to EEOC Order 205.001, Appendix B, Attachment 4, § a(5), this Notice will remain in effect until rescinded or superseded.

5. *ORIGINATOR:* Title VII/EPA/ADEA Division, Office of Legal Counsel.

6. *INSTRUCTIONS:* File after Section 615 of Volume II of the Compliance Manual.

6/18/99   /s/
Date   Ida L. Castro
               Chairwoman

Table of Contents

Enforcement Guidance on Vicarious Employer Liability for Unlawful Harassment by Supervisors

I. Introduction

In *Burlington Industries, Inc. v. Ellerth,* 118 S. Ct. 2257 (1998), and *Faragher v. City of Boca Raton,* 118 S. Ct. 2275 (1998), the Supreme Court made clear that employers are subject to vicarious liability for unlawful harassment by supervisors. The standard of liability set forth in these decisions is premised on two principles: 1) an employer is responsible for the acts of its supervisors, and 2) employers should be encouraged to prevent harassment and employees should be encouraged to avoid or limit the harm from harassment. In order to accommodate these principles, the Court held that an employer is always liable for a supervisor's harassment if it culminates in a tangible employment action. However, if it does not, the employer may be able to avoid liability or limit damages by establishing an affirmative defense that includes two necessary elements:

(a) the employer exercised reasonable care to prevent and correct promptly any harassing behavior, and

(b) the employee unreasonably failed to take advantage of any preventive or corrective opportunities provided by the employer or to avoid harm otherwise.

While the *Faragher* and *Ellerth* decisions addressed sexual harassment, the Court's analysis drew upon standards set forth in cases involving harassment on other protected bases. Moreover, the Commission has always taken the position that the same basic standards apply to all types of prohibited harassment.[1] Thus, the standard of liability set forth in the decisions applies to all forms of unlawful harassment. (See section II, below.)

Harassment remains a pervasive problem in American workplaces. The number of harassment charges filed with the EEOC and state fair employment practices agencies has risen significantly in recent years. For example, the number of sexual harassment charges has increased from 6,883 in fiscal year 1991 to 15,618 in fiscal year 1998. The number of racial harassment charges rose from 4,910 to 9,908 charges in the same time period.

While the anti-discrimination statutes seek to remedy discrimination, their primary purpose is to prevent violations. The Supreme Court, in *Faragher* and *Ellerth,* relied on Commission guidance which has long advised employers to take all necessary steps to prevent harassment.[2] The new affirmative defense gives credit for such preventive efforts by an employer, thereby "implement[ing] clear statutory policy and complement[ing] the Government's Title VII enforcement efforts."[3]

The question of liability arises only after there is a determination that unlawful harassment occurred. Harassment does not violate federal law unless it involves discriminatory treatment on the basis of race, color, sex, religion, national origin, age of 40 or older, disability, or protected activity under the anti-discrimination statutes. Furthermore, the anti-discrimination statutes are not a "general civility code."[4] Thus federal law does not prohibit simple teasing, offhand comments, or isolated incidents that are not "extremely serious."[5] Rather, the conduct must be "so objectively offensive as to alter the 'conditions' of the victim's employment."[6] The conditions of employment are altered only if the harassment culminated in a tangible employment action or was sufficiently severe or pervasive to create a hostile work environment.[7] Existing Commission guidance on the standards for determining whether challenged conduct rises to the level of unlawful harassment remains in effect.

This document supersedes previous Commission guidance on the issue of vicarious liability for harassment by supervisors.[8] The Commission's long-standing guidance on employer liability for harassment by co-workers remains in effect—an employer is liable if it knew or should have known of the misconduct, unless it can show that it took immediate and appropriate corrective action.[9] The standard is the same in the case of non-employees, but the employer's control over such individuals' misconduct is considered.[10]

II. The Vicarious Liability Rule Applies to Unlawful Harassment on All Covered Bases

The rule in *Ellerth* and *Faragher* regarding vicarious liability applies to harassment by supervisors based on race, color, sex (whether or not of a sexual nature[11]), religion, national origin, protected activity,[12] age, or disability.[13] Thus, employers should establish anti-harassment policies and complaint procedures covering *all* forms of unlawful harassment.[14]

III. Who Qualifies as a Supervisor?

A. Harasser in Supervisory Chain of Command

An employer is subject to vicarious liability for unlawful harassment if the harassment was committed by "a supervisor with immediate (or successively higher) authority over the employee."[15] Thus, it is critical to determine whether the person who engaged in unlawful harassment had supervisory authority over the complainant.

The federal employment discrimination statutes do not contain or define the term "supervisor."[16] The statutes make employers liable for the discriminatory acts of their "agents,"[17] and supervisors are agents of their employers. However, agency principles "may not be transferable in all their particulars" to the federal employment discrimination statutes.[18] The determination of whether an individual has sufficient authority to qualify as a "supervisor" for purposes of vicarious liability cannot be resolved by a purely mechanical application of agency law.[19] Rather, the purposes of the anti-discrimination statutes and the reasoning of the Supreme Court decisions on harassment must be considered.

The Supreme Court, in *Faragher* and *Ellerth,* reasoned that vicarious liability for supervisor harassment is appropriate because supervisors are aided in such misconduct by the authority that the employers delegated to them.[20] Therefore, that authority must be of a sufficient magnitude so as to assist the harasser explicitly or implicitly in carrying out the harassment. The determination as to whether a harasser had such authority is based on his or her job function rather than job title (*e.g.,* "team leader") and must be based on the specific facts.

An individual qualifies as an employee's "supervisor" if:

a. the individual has authority to undertake or recommend tangible employment decisions affecting the employee; or

b. the individual has authority to direct the employee's daily work activities.

1. Authority to Undertake or Recommend Tangible Employment Actions

An individual qualifies as an employee's "supervisor" if he or she is authorized to undertake tangible employment decisions affecting the employee. "Tangible employment decisions" are decisions that significantly change another employee's employment status. (For a detailed explanation of what constitutes a tangible employment action, see subsection IV(B), below.) Such actions include, but are not limited to, hiring, firing, promoting, demoting, and reassigning the employee. As the Supreme Court stated, "[t]angible employment actions fall within the special province of the supervisor."[21]

An individual whose job responsibilities include the authority to recommend tangible job decisions affecting an employee qualifies as his or her supervisor even if the individual does not have the final say. As the Supreme Court recognized in *Ellerth,* a tangible employment decision "may be subject to review by higher level supervisors."[22] As long as the individual's recommendation is given substantial weight by the final decisionmaker(s), that individual meets the definition of supervisor.

2. Authority to Direct Employee's Daily Work Activities

An individual who is authorized to direct another employee's day-to-day work activities qualifies as his or her supervisor even if that individual does not have the authority to undertake or recommend tangible job decisions. Such an individual's ability to commit harassment is enhanced by his or her authority to increase the employee's workload or assign undesirable tasks, and hence it is appropriate to consider such a person a "supervisor" when determining whether the employer is vicariously liable.

In *Faragher,* one of the harassers was authorized to hire, supervise, counsel, and discipline lifeguards, while the other harasser was responsible for making the lifeguards' daily work assignments and supervising their work and fitness training.[23] There was no question that the Court viewed them *both* as "supervisors," even though one of them apparently lacked authority regarding tangible job decisions.[24]

An individual who is temporarily authorized to direct another employee's daily work activities qualifies as his or her "supervisor" during that time period. Accordingly, the employer would be subject to vicarious liability if that individual commits unlawful harassment of a subordinate while serving as his or her supervisor.

On the other hand, someone who merely relays other officials' instructions regarding work assignments and reports back to those officials does not have true supervisory authority. Furthermore, someone who directs only a limited number of tasks or assignments would not qualify as a "supervisor." For example, an individual whose delegated authority is confined to coordinating a work project of limited scope is not a "supervisor."

B. Harasser Outside Supervisory Chain of Command
In some circumstances, an employer may be subject to vicarious liability for harassment by a supervisor who does not have actual authority over the employee. Such a result is appropriate if the employee reasonably believed that the harasser had such power.[25] The employee might have such a belief because, for example, the chains of command are unclear. Alternatively, the employee might reasonably believe that a harasser with broad delegated powers has the ability to significantly influence employment decisions affecting him or her even if the harasser is outside the employee's chain of command.

If the harasser had no actual supervisory power over the employee, and the employee did not reasonably believe that the harasser had such authority, then the standard of liability for co-worker harassment applies.

IV. Harassment by Supervisor That Results in a Tangible Employment Action
A. Standard of Liability
An employer is always liable for harassment by a supervisor on a prohibited basis that culminates in a tangible employment action. No affirmative defense is available in such cases.[26] The Supreme Court recognized that this result is appropriate because an employer acts through its supervisors, and a supervisor's undertaking of a tangible employment action constitutes an act of the employer.[27]

B. Definition of "Tangible Employment Action"
A tangible employment action is "a significant change in employment status."[28] Unfulfilled threats are insufficient. Characteristics of a tangible employment action are:[29]

1. A tangible employment action is the means by which the supervisor brings the official power of the enterprise to bear on subordinates, as demonstrated by the following:
   - it requires an official act of the enterprise;
   - it usually is documented in official company records;
   - it may be subject to review by higher level supervisors; and
   - it often requires the formal approval of the enterprise and use of its internal processes.
2. A tangible employment action usually inflicts direct economic harm.
3. A tangible employment action, in most instances, can only be caused by a supervisor or other person acting with the authority of the company.

Examples of tangible employment actions include:[30]

- hiring and firing;
- promotion and failure to promote;
- demotion;[31]
- undesirable reassignment;
- a decision causing a significant change in benefits;
- compensation decisions; and
- work assignment.

Any employment action qualifies as "tangible" if it results in a significant change in employment status. For example, significantly changing an individual's duties in his or her existing job constitutes a tangible employment action regardless of whether the individual retains the same salary and benefits.[32] Similarly, altering an individual's duties in a way that blocks his or her opportunity for promotion or salary increases also constitutes a tangible employment action.[33]

On the other hand, an employment action does not reach the threshold of "tangible" if it results in only an insignificant change in the complainant's employment status. For example, altering an individual's job title does not qualify as a tangible employment action if there is no change in salary, benefits, duties, or prestige, and the only effect is a bruised ego.[34] However, if there is a significant change in the status of the position because the new title is less prestigious and thereby effectively constitutes a demotion, a tangible employment action would be found.[35]

If a supervisor undertakes or recommends a tangible job action based on a subordinate's response to unwelcome sexual demands, the employer is liable and cannot raise the affirmative defense. The result is the same whether the employee rejects the demands and is subjected to an adverse tangible employment action or submits to the demands and consequently obtains a tangible job benefit.[36] Such harassment previously would have been characterized as "quid pro quo." It would be a perverse result if the employer is fore-closed from raising the affirmative defense if its supervisor denies a tangible job benefit based on an employee's rejection of unwelcome sexual demands, but can raise the defense if its supervisor grants a tangible job benefit based on submission to such demands. The Commission rejects such an analysis. In both those situations the supervisor undertakes a tangible employment action on a discriminatory basis. The Supreme Court stated that there must be a significant *change* in employment status; it did not require that the change be adverse in order to qualify as tangible.[37]

If a challenged employment action is not "tangible," it may still be consid-ered, along with other evidence, as part of a hostile environment claim that is subject to the affirmative defense. In *Ellerth*, the Court concluded that there was no tangible employment action because the supervisor never car-ried out his threats of job harm. Ellerth could still proceed with her claim of harassment, but the claim was properly "categorized as a hostile work envi-ronment claim which requires a showing of severe or pervasive conduct." 118 S. Ct. at 2265.

C. Link Between Harassment and Tangible Employment Action
When harassment culminates in a tangible employment action, the employer cannot raise the affirmative defense. This sort of claim is analyzed like any other case in which a challenged employment action is alleged to be discrim-inatory. If the employer produces evidence of a non-discriminatory explana-tion for the tangible employment action, a determination must be made whether that explanation is a pretext designed to hide a discriminatory motive.

For example, if an employee alleged that she was demoted because she refused her supervisor's sexual advances, a determination would have to be made whether the demotion was *because* of her response to the advances, and hence because of her sex. Similarly, if an employee alleges that he was discharged after being subjected to severe or pervasive harassment by his supervisor based on his national origin, a determination would have to be made whether the discharge was *because* of the employee's national origin.

A strong inference of discrimination will arise whenever a harassing super-visor undertakes or has significant input into a tangible employment action affecting the victim,[38] because it can be "assume[d] that the harasser . . . could not act as an objective, non-discriminatory decisionmaker with respect to the plaintiff."[39] However, if the employer produces evidence of a non-discriminatory reason for the action, the employee will have to prove that the asserted reason was a pretext designed to hide the true discriminatory motive.

If it is determined that the tangible action was based on a discriminatory reason linked to the preceding harassment, relief could be sought for the en-tire pattern of misconduct culminating in the tangible employment action, and no affirmative defense is available.[40] However, the harassment preceding the tangible employment action must be severe or pervasive in order to be actionable.[41] If the tangible employment action was based on a non-discriminatory motive, then the employer would have an opportunity to raise the affirmative defense to a claim based on the preceding harassment.[42]

V. Harassment by Supervisor That Does Not Result in a Tangible Employment Action
A. Standard of Liability

When harassment by a supervisor creates an unlawful hostile environment but does not result in a tangible employment action, the employer can raise an affirmative defense to liability or damages, which it must prove by a preponderance of the evidence. The defense consists of two necessary elements:

(a) the employer exercised reasonable care to prevent and correct promptly any harassment; and

(b) the employee unreasonably failed to take advantage of any preventive or corrective opportunities provided by the employer or to avoid harm otherwise.

B. Effect of Standard

If an employer can prove that it discharged its duty of reasonable care and that the employee could have avoided all of the harm but unreasonably failed to do so, the employer will avoid all liability for unlawful harassment.[43] For example, if an employee was subjected to a pattern of disability-based harassment that created an unlawful hostile environment, but the employee unreasonably failed to complain to management before she suffered emotional harm and the employer exercised reasonable care to prevent and promptly correct the harassment, then the employer will avoid all liability.

If an employer cannot prove that it discharged its duty of reasonable care *and* that the employee unreasonably failed to avoid the harm, the employer will be liable. For example, if unlawful harassment by a supervisor occurred and the employer failed to exercise reasonable care to prevent it, the employer will be liable even if the employee unreasonably failed to complain to management or even if the employer took prompt and appropriate corrective action when it gained notice.[44]

In most circumstances, if employers and employees discharge their respective duties of reasonable care, unlawful harassment will be prevented and there will be no reason to consider questions of liability. An effective complaint procedure "encourages employees to report harassing conduct before it becomes severe or pervasive,"[45] and if an employee promptly utilizes that procedure, the employer can usually stop the harassment before actionable harm occurs.[46]

In some circumstances, however, unlawful harassment will occur and harm will result despite the exercise of requisite legal care by the employer and employee. For example, if an employee's supervisor directed frequent, egregious racial epithets at him that caused emotional harm virtually from the outset, and the employee promptly complained, corrective action by the employer could prevent further harm but might not correct the actionable harm that the employee already had suffered.[47] Alternatively, if an employee complained about harassment before it became severe or pervasive, remedial measures undertaken by the employer might fail to stop the harassment before it reaches an actionable level, even if those measures are reasonably calculated to halt it. In these circumstances, the employer will be liable because the defense requires proof that it exercised reasonable legal care *and* that the employee unreasonably failed to avoid the harm. While a notice-based negligence standard would absolve the employer of liability, the standard set forth in *Ellerth* and *Faragher* does not. As the Court explained, vicarious liability sets a "more stringent standard" for the employer than the "minimum standard" of negligence theory.[48]

While this result may seem harsh to a law-abiding employer, it is consistent with liability standards under the anti-discrimination statutes which generally make employers responsible for the discriminatory acts of their supervisors.[49] If, for example, a supervisor rejects a candidate for promotion because of national origin–based bias, the employer will be liable regardless of whether the employee complained to higher management and regardless of whether higher management had any knowledge about the supervisor's motivation.[50] Harassment is the only type of discrimination carried out by a supervisor for which an employer can avoid liability, and that limitation must be construed narrowly. The employer will be shielded from liability for harassment by a supervisor only if it proves that it exercised reasonable care in preventing and correcting the harassment *and* that the employee unreasonably failed to avoid all of the harm. If both parties exercise reasonable care, the defense will fail.

In some cases, an employer will be unable to avoid liability completely, but may be able to establish the affirmative defense as a means to limit damages.[51] The defense only limits damages where the employee reasonably could have avoided some but not all of the harm from the harassment. In the example above, in which the supervisor used frequent, egregious racial epithets, an unreasonable delay by the employee in complaining could limit damages but not eliminate liability entirely. This is because a reasonably prompt complaint would have reduced, but not eliminated, the actionable harm.[52]

C. First Prong of Affirmative Defense: Employer's Duty to Exercise Reasonable Care

The first prong of the affirmative defense requires a showing by the employer that it undertook reasonable care to prevent and promptly correct harassment. Such reasonable care generally requires an employer to establish, disseminate, and enforce an anti-harassment policy and complaint procedure and to take other reasonable steps to prevent and correct harassment. The steps described below are not mandatory requirements—whether or not an employer can prove that it exercised reasonable care depends on the particular factual circumstances and, in some cases, the nature of the employer's workforce. Small employers may be able to effectively prevent and correct harassment through informal means, while larger employers may have to institute more formal mechanisms.[53]

There are no "safe harbors" for employers based on the written content of policies and procedures. Even the best policy and complaint procedure will not alone satisfy the burden of proving reasonable care if, in the particular circumstances of a claim, the employer failed to implement its process effectively.[54] If, for example, the employer has an adequate policy and complaint procedure and properly responded to an employee's complaint of harassment, but management ignored previous complaints by other employees about the same harasser, then the employer has not exercised reasonable care in preventing the harassment.[55] Similarly, if the employer has an adequate policy and complaint procedure but an official failed to carry out his or her responsibility to conduct an effective investigation of a harassment complaint, the employer has not discharged its duty to exercise reasonable care. Alternatively, lack of a formal policy and complaint procedure will not defeat the defense if the employer exercised sufficient care through other means.

1. Policy and Complaint Procedure

It generally is necessary for employers to establish, publicize, and enforce anti-harassment policies and complaint procedures. As the Supreme Court stated, "Title VII is designed to encourage the creation of anti-harassment policies and effective grievance mechanisms." *Ellerth*, 118 S. Ct. at 2270. While the Court noted that this "is not necessary in every instance as a matter of law,"[56] failure to do so will make it difficult for an employer to prove that it exercised reasonable care to prevent and correct harassment.[57] (See section V(C)(3), below, for discussion of preventive and corrective measures by small businesses.)

An employer should provide every employee with a copy of the policy and complaint procedure, and redistribute it periodically. The policy and complaint procedure should be written in a way that will be understood by all employees in the employer's workforce. Other measures to ensure effective dissemination of the policy and complaint procedure include posting them in central locations and incorporating them into employee handbooks. If feasible, the employer should provide training to all employees to ensure that they understand their rights and responsibilities.

An anti-harassment policy and complaint procedure should contain, at a minimum, the following elements:

- A clear explanation of prohibited conduct;
- Assurance that employees who make complaints of harassment or provide information related to such complaints will be protected against retaliation;

- A clearly described complaint process that provides accessible avenues of complaint;
- Assurance that the employer will protect the confidentiality of harassment complaints to the extent possible;
- A complaint process that provides a prompt, thorough, and impartial investigation; and
- Assurance that the employer will take immediate and appropriate corrective action when it determines that harassment has occurred.

The above elements are explained in the following subsections.

a. Prohibition Against Harassment

An employer's policy should make clear that it will not tolerate harassment based on sex (with or without sexual conduct), race, color, religion, national origin, age, disability, and protected activity (*i.e.,* opposition to prohibited discrimination or participation in the statutory complaint process). This prohibition should cover harassment by *anyone* in the workplace—supervisors, co-workers, or non-employees.[58] Management should convey the seriousness of the prohibition. One way to do that is for the mandate to "come from the top," *i.e.,* from upper management.

The policy should encourage employees to report harassment *before* it becomes severe or pervasive. While isolated incidents of harassment generally do not violate federal law, a pattern of such incidents may be unlawful. Therefore, to discharge its duty of preventive care, the employer must make clear to employees that it will stop harassment before it rises to the level of a violation of federal law.

b. Protection Against Retaliation

An employer should make clear that it will not tolerate adverse treatment of employees because they report harassment or provide information related to such complaints. An anti-harassment policy and complaint procedure will not be effective without such an assurance.[59]

Management should undertake whatever measures are necessary to ensure that retaliation does not occur. For example, when management investigates a complaint of harassment, the official who interviews the parties and witnesses should remind these individuals about the prohibition against retaliation. Management also should scrutinize employment decisions affecting the complainant and witnesses during and after the investigation to ensure that such decisions are not based on retaliatory motives.

c. Effective Complaint Process

An employer's harassment complaint procedure should be designed to encourage victims to come forward. To that end, it should clearly explain the process and ensure that there are no unreasonable obstacles to complaints. A complaint procedure should not be rigid, since that could defeat the goal of preventing and correcting harassment. When an employee complains to management about alleged harassment, the employer is obligated to investigate the allegation regardless of whether it conforms to a particular format or is made in writing.

The complaint procedure should provide accessible points of contact for the initial complaint.[60] A complaint process is not effective if employees are always required to complain first to their supervisors about alleged harassment, since the supervisor may be a harasser.[61] Moreover, reasonable care in preventing and correcting harassment requires an employer to instruct all supervisors to report complaints of harassment to appropriate officials.[62]

It is advisable for an employer to designate at least one official outside an employee's chain of command to take complaints of harassment. For example, if the employer has an office of human resources, one or more officials in that office could be authorized to take complaints. Allowing an employee to bypass his or her chain of command provides additional assurance that the complaint will be handled in an impartial manner, since an employee who reports harassment by his or her supervisor may feel that officials within the chain of command will more readily believe the supervisor's version of events.

It also is important for an employer's anti-harassment policy and complaint procedure to contain information about the time frames for filing charges of unlawful harassment with the EEOC or state fair employment practice agencies and to explain that the deadline runs from the last date of unlawful harassment, not from the date that the complaint to the employer is resolved.[63] While a prompt complaint process should make it feasible for an employee to delay deciding whether to file a charge until the complaint to the employer is resolved, he or she is not required to do so.[64]

d. Confidentiality

An employer should make clear to employees that it will protect the confidentiality of harassment allegations to the extent possible. An employer cannot guarantee complete confidentiality, since it cannot conduct an effective investigation without revealing certain information to the alleged harasser and potential witnesses. However, information about the allegation of harassment should be shared only with those who need to know about it. Records relating to harassment complaints should be kept confidential on the same basis.[65]

A conflict between an employee's desire for confidentiality and the employer's duty to investigate may arise if an employee informs a supervisor about alleged harassment, but asks him or her to keep the matter confidential and take no action. Inaction by the supervisor in such circumstances could lead to employer liability. While it may seem reasonable to let the employee determine whether to pursue a complaint, the employer must discharge its duty to prevent and correct harassment.[66] One mechanism to help avoid such conflicts would be for the employer to set up an informational phone line which employees can use to discuss questions or concerns about harassment on an anonymous basis.[67]

e. Effective Investigative Process

An employer should set up a mechanism for a prompt, thorough, and impartial investigation into alleged harassment. As soon as management learns about alleged harassment, it should determine whether a detailed fact-finding investigation is necessary. For example, if the alleged harasser does not deny the accusation, there would be no need to interview witnesses, and the employer could immediately determine appropriate corrective action.

If a fact-finding investigation is necessary, it should be launched immediately. The amount of time that it will take to complete the investigation will depend on the particular circumstances.[68] If, for example, multiple individuals were allegedly harassed, then it will take longer to interview the parties and witnesses.

It may be necessary to undertake intermediate measures before completing the investigation to ensure that further harassment does not occur. Examples of such measures are making scheduling changes so as to avoid contact between the parties; transferring the alleged harasser; or placing the alleged harasser on non-disciplinary leave with pay pending the conclusion of the investigation. The complainant should not be involuntarily transferred or otherwise burdened, since such measures could constitute unlawful retaliation.

The employer should ensure that the individual who conducts the investigation will objectively gather and consider the relevant facts. The alleged harasser should not have supervisory authority over the individual who conducts the investigation and should not have any direct or indirect control over the investigation. Whoever conducts the investigation should be well-trained in the skills that are required for interviewing witnesses and evaluating credibility.

i. Questions to Ask Parties and Witnesses

When detailed fact-finding is necessary, the investigator should interview the complainant, the alleged harasser, and third parties who could reasonably be expected to have relevant information. Information relating to the personal lives of the parties outside the workplace would be relevant only in unusual circumstances. When interviewing the parties and witnesses, the investigator should refrain from offering his or her opinion.

The following are examples of questions that may be appropriate to ask the parties and potential witnesses. Any actual investigation must be tailored to the particular facts.

Questions to Ask the Complainant:

- Who, what, when, where, and how: *Who* committed the alleged harassment? *What* exactly occurred or was said? *When* did it occur and is it still ongoing? *Where* did it occur? *How often* did it occur? *How* did it affect you?
- How did you react? What response did you make when the incident(s) occurred or afterwards?
- How did the harassment affect you? Has your job been affected in any way?
- Are there any persons who have relevant information? Was anyone present when the alleged harassment occurred? Did you tell anyone about it? Did anyone see you immediately after episodes of alleged harassment?
- Did the person who harassed you harass anyone else? Do you know whether anyone complained about harassment by that person?
- Are there any notes, physical evidence, or other documentation regarding the incident(s)?
- How would you like to see the situation resolved?
- Do you know of any other relevant information?

Questions to Ask the Alleged Harasser:

- What is your response to the allegations?
- If the harasser claims that the allegations are false, ask why the complainant might lie.
- Are there any persons who have relevant information?
- Are there any notes, physical evidence, or other documentation regarding the incident(s)?
- Do you know of any other relevant information?

Questions to Ask Third Parties:

- What did you see or hear? When did this occur? Describe the alleged harasser's behavior toward the complainant and toward others in the workplace.
- What did the complainant tell you? When did s/he tell you this?
- Do you know of any other relevant information?
- Are there other persons who have relevant information?

ii. Credibility Determinations

If there are conflicting versions of relevant events, the employer will have to weigh each party's credibility. Credibility assessments can be critical in determining whether the alleged harassment in fact occurred. Factors to consider include:

- **Inherent plausibility:** Is the testimony believable on its face? Does it make sense?
- **Demeanor:** Did the person seem to be telling the truth or lying?

- **Motive to falsify:** Did the person have a reason to lie?
- **Corroboration:** Is there witness testimony (such as testimony by eye-witnesses, people who saw the person soon after the alleged incidents, or people who discussed the incidents with him or her at around the time that they occurred) or physical evidence (such as written documentation) that corroborates the party's testimony?
- **Past record:** Did the alleged harasser have a history of similar behavior in the past?

None of the above factors are determinative as to credibility. For example, the fact that there are no eye-witnesses to the alleged harassment by no means necessarily defeats the complainant's credibility, since harassment often occurs behind closed doors. Furthermore, the fact that the alleged harasser engaged in similar behavior in the past does not necessarily mean that he or she did so again.

iii. Reaching a Determination

Once all of the evidence is in, interviews are finalized, and credibility issues are resolved, management should make a determination as to whether harassment occurred. That determination could be made by the investigator, or by a management official who reviews the investigator's report. The parties should be informed of the determination.

In some circumstances, it may be difficult for management to reach a determination because of direct contradictions between the parties and a lack of documentary or eye-witness corroboration. In such cases, a credibility assessment may form the basis for a determination, based on factors such as those set forth above.

If no determination can be made because the evidence is inconclusive, the employer should still undertake further preventive measures, such as training and monitoring.

f. Assurance of Immediate and Appropriate Corrective Action

An employer should make clear that it will undertake immediate and appropriate corrective action, including discipline, whenever it determines that harassment has occurred in violation of the employer's policy. Management should inform both parties about these measures.[69]

Remedial measures should be designed to stop the harassment, correct its effects on the employee, and ensure that the harassment does not recur. These remedial measures need not be those that the employee requests or prefers, as long as they are effective.

In determining disciplinary measures, management should keep in mind that the employer could be found liable if the harassment does not stop. At the same time, management may have concerns that overly punitive measures may subject the employer to claims such as wrongful discharge, and may simply be inappropriate.

To balance the competing concerns, disciplinary measures should be proportional to the seriousness of the offense.[70] If the harassment was minor, such as a small number of "off-color" remarks by an individual with no prior history of similar misconduct, then counseling and an oral warning might be all that is necessary. On the other hand, if the harassment was severe or persistent, then suspension or discharge may be appropriate.[71]

Remedial measures should not adversely affect the complainant. Thus, for example, if it is necessary to separate the parties, then the harasser should be transferred (unless the complainant prefers otherwise).[72] Remedial responses that penalize the complainant could constitute unlawful retaliation and are not effective in correcting the harassment.[73]

Remedial measures also should correct the effects of the harassment. Such measures should be designed to put the employee in the position s/he would have been in had the misconduct not occurred.

Examples of Measures to Stop the Harassment and Ensure That It Does Not Recur:
- oral[74] or written warning or reprimand;
- transfer or reassignment;
- demotion;
- reduction of wages;
- suspension;
- discharge;
- training or counseling of harasser to ensure that s/he understands why his or her conduct violated the employer's anti-harassment policy; and
- monitoring of harasser to ensure that harassment stops.

Examples of Measures to Correct the Effects of the Harassment:
- restoration of leave taken because of the harassment;
- expungement of negative evaluation(s) in employee's personnel file that arose from the harassment;
- reinstatement;
- apology by the harasser;
- monitoring treatment of employee to ensure that s/he is not subjected to retaliation by the harasser or others in the workplace because of the complaint; and
- correction of any other harm caused by the harassment (e.g., compensation for losses).

2. Other Preventive and Corrective Measures

An employer's responsibility to exercise reasonable care to prevent and correct harassment is not limited to implementing an anti-harassment policy and complaint procedure. As the Supreme Court stated, "the employer has a greater opportunity to guard against misconduct by supervisors than by common workers; employers have greater opportunity and incentive to screen them, train them, and monitor their performance." *Faragher,* 118 S. Ct. at 2291.

An employer's duty to exercise due care includes instructing all of its supervisors and managers to address or report to appropriate officials complaints of harassment regardless of whether they are officially designated to take complaints[75] and regardless of whether a complaint was framed in a way that conforms to the organization's particular complaint procedures.[76] For example, if an employee files an EEOC charge alleging unlawful harassment, the employer should launch an internal investigation even if the employee did not complain to management through its internal complaint process.

Furthermore, due care requires management to correct harassment regardless of whether an employee files an internal complaint, if the conduct is clearly unwelcome. For example, if there are areas in the workplace with graffiti containing racial or sexual epithets, management should eliminate the graffiti and not wait for an internal complaint.[77]

An employer should ensure that its supervisors and managers understand their responsibilities under the organization's anti-harassment policy and complaint procedure. Periodic training of those individuals can help achieve that result. Such training should explain the types of conduct that violate the employer's anti-harassment policy; the seriousness of the policy; the responsibilities of supervisors and managers when they learn of alleged harassment; and the prohibition against retaliation.

An employer should keep track of its supervisors' and managers' conduct to make sure that they carry out their responsibilities under the organization's anti-harassment program.[78] For example, an employer could include such compliance in formal evaluations.

Reasonable preventive measures include screening applicants for supervisory jobs to see if any have a record of engaging in harassment. If so, it may be necessary for the employer to reject a candidate on that basis or to take additional steps to prevent harassment by that individual.

Finally, it is advisable for an employer to keep records of all complaints of harassment. Without such records, the employer could be unaware of a pattern of harassment by the same individual. Such a pattern would be relevant to credibility assessments and disciplinary measures.[79]

3. Small Businesses

It may not be necessary for an employer of a small workforce to implement the type of formal complaint process described above. If it puts into place an effective, informal mechanism to prevent and correct harassment, a small employer could still satisfy the first prong of the affirmative defense to a claim of harassment.[80] As the Court recognized in *Faragher,* an employer of a small workforce might informally exercise sufficient care to prevent harassment.[81]

For example, such an employer's failure to disseminate a written policy against harassment on protected bases would not undermine the affirmative defense if it effectively communicated the prohibition and an effective complaint procedure to all employees at staff meetings. An owner of a small business who regularly meets with all of his or her employees might tell them at monthly staff meetings that he or she will not tolerate harassment and that anyone who experiences harassment should bring it "straight to the top."

If a complaint is made, the business, like any other employer, must conduct a prompt, thorough, and impartial investigation and undertake swift and appropriate corrective action where appropriate. The questions set forth in Section V(C)(1)(e)(i), above, can help guide the inquiry and the factors set forth in Section V(C)(1)(e)(ii) should be considered in evaluating the credibility of each of the parties.

D. Second Prong of Affirmative Defense: Employee's Duty to Exercise Reasonable Care

The second prong of the affirmative defense requires a showing by the employer that the aggrieved employee "unreasonably failed to take advantage of any preventive or corrective opportunities provided by the employer or to avoid harm otherwise." *Faragher,* 118 S. Ct. at 2293; *Ellerth,* 118 S. Ct. at 2270.

This element of the defense arises from the general theory "that a victim has a duty 'to use such means as are reasonable under the circumstances to avoid or minimize the damages' that result from violations of the statute." *Faragher,* 18 S. Ct. at 2292, *quoting Ford Motor Co. v. EEOC,* 458 U.S. 219, 231 n.15 (1982). Thus an employer who exercised reasonable care as described in subsection V(C), above, is not liable for unlawful harassment if the aggrieved employee could have avoided all of the actionable harm. If some but not all of the harm could have been avoided, then an award of damages will be mitigated accordingly.[82]

A complaint by an employee does not automatically defeat the employer's affirmative defense. If, for example, the employee provided no information to support his or her allegation, gave untruthful information, or otherwise failed to cooperate in the investigation, the complaint would not qualify as an effort to avoid harm. Furthermore, if the employee unreasonably delayed complaining, and an earlier complaint could have reduced the harm, then the affirmative defense could operate to reduce damages.

Proof that the employee unreasonably failed to use any complaint procedure provided by the employer will normally satisfy the employer's burden.[83] However, it is important to emphasize that an employee who failed to complain does not carry a burden of proving the reasonableness of that decision. Rather, the burden lies with the employer to prove that the employee's failure to complain was unreasonable.

1. Failure to Complain

A determination as to whether an employee unreasonably failed to complain or otherwise avoid harm depends on the particular circumstances and information available to the employee *at that time*.[84] An employee should not necessarily be expected to complain to management immediately after the first or second incident of relatively minor harassment. Workplaces need not become battlegrounds where every minor, unwelcome remark based on race, sex, or another protected category triggers a complaint and investigation. An employee might reasonably ignore a small number of incidents, hoping that the harassment will stop without resort to the complaint process.[85] The employee may directly say to the harasser that s/he wants the misconduct to stop, and then wait to see if that is effective in ending the harassment before complaining to management. If the harassment persists, however, then further delay in complaining might be found unreasonable.

There might be other reasonable explanations for an employee's delay in complaining or entire failure to utilize the employer's complaint process. For example, the employee might have had reason to believe that:[86]

- using the complaint mechanism entailed a risk of retaliation;
- there were obstacles to complaints; and
- the complaint mechanism was not effective.

To establish the second prong of the affirmative defense, the employer must prove that the belief or perception underlying the employee's failure to complain was unreasonable.

a. Risk of Retaliation

An employer cannot establish that an employee unreasonably failed to use its complaint procedure if that employee reasonably feared retaliation. Surveys have shown that employees who are subjected to harassment frequently do not complain to management due to fear of retaliation.[87] To assure employees that such a fear is unwarranted, the employer must clearly communicate and enforce a policy that no employee will be retaliated against for complaining of harassment.

b. Obstacles to Complaints

An employee's failure to use the employer's complaint procedure would be reasonable if that failure was based on unnecessary obstacles to complaints. For example, if the process entailed undue expense by the employee,[88] inaccessible points of contact for making complaints,[89] or unnecessarily intimidating or burdensome requirements, failure to invoke it on such a basis would be reasonable.

An employee's failure to participate in a mandatory mediation or other alternative dispute resolution process also does not constitute unreasonable failure to avoid harm. While an employee can be expected to cooperate in the employer's investigation by providing relevant information, an employee can never be required to waive rights, either substantive or procedural, as an element of his or her exercise of reasonable care.[90] Nor must an employee have to try to resolve the matter with the harasser as an element of exercising due care.

c. Perception That Complaint Process Was Ineffective

An employer cannot establish the second prong of the defense based on the employee's failure to complain if that failure was based on a reasonable belief that the process was ineffective. For example, an employee would have a reasonable basis to believe that the complaint process is ineffective if the procedure required the employee to complain initially to the harassing supervisor. Such a reasonable basis also would be found if he or she was aware of instances in which co-workers' complaints failed to stop harassment. One way to increase employees' confidence in the efficacy of the complaint process would be for the employer to release general information to employees about corrective and disciplinary measures undertaken to stop harassment.[91]

2. Other Efforts to Avoid Harm

Generally, an employer can prove the second prong of the affirmative defense if the employee unreasonably failed to utilize its complaint process. However, such proof will not establish the defense if the employee made other efforts to avoid harm.

For example, a prompt complaint by the employee to the EEOC or a state fair employment practices agency while the harassment is ongoing could qualify as such an effort. A union grievance could also qualify as an effort to avoid harm.[92] Similarly, a staffing firm worker who is harassed at the client's workplace might report the harassment either to the staffing firm or to the client, reasonably expecting that either would act to correct the problem.[93] Thus the worker's failure to complain to one of those entities would not bar him or her from subsequently bringing a claim against it.

With these and any other efforts to avoid harm, the timing of the complaint could affect liability or damages. If the employee could have avoided some of the harm by complaining earlier, then damages would be mitigated accordingly.

VI. Harassment by "Alter Ego" of Employer

A. Standard of Liability

An employer is liable for unlawful harassment whenever the harasser is of a sufficiently high rank to fall "within that class . . . who may be treated as the organization's proxy." *Faragher,* 118 S. Ct. at 2284.[94] In such circumstances, the official's unlawful harassment is imputed automatically to the employer.[95] Thus the employer cannot raise the affirmative defense, even if the harassment did not result in a tangible employment action.

B. Officials Who Qualify as "Alter Egos" or "Proxies"

The Court, in *Faragher,* cited the following examples of officials whose harassment could be imputed automatically to the employer:

- president[96]
- owner[97]
- partner[98]
- corporate officer

Faragher, 118 S. Ct. at 2284.

VII. Conclusion

The Supreme Court's rulings in *Ellerth* and *Faragher* create an incentive for employers to implement and enforce strong policies prohibiting harassment and effective complaint procedures. The rulings also create an incentive for employees to alert management about harassment before it becomes severe and pervasive. If employers and employees undertake these steps, unlawful harassment can often be prevented, thereby effectuating an important goal of the anti-discrimination statutes.

Notes

1. *See, e.g.,* 29 C.F.R. § 1604.11 n. 1 ("The principles involved here continue to apply to race, color, religion or national origin."); EEOC Compliance Manual Section 615.11(a) (BNA) 615:0025 ("Title VII law and agency principles will guide the determination of whether an employer is liable for age harassment by its supervisors, employees, or non-employees").

2. *See* 1980 Guidelines at 29 C.F.R. § 1604.11(f) and Policy Guidance on Current Issues of Sexual Harassment, Section E, 8 FEP Manual 405:6699 (Mar. 19, 1990), *quoted in Faragher,* 118 S. Ct. at 2292.

3. *Faragher,* 118 S. Ct. at 2292.

4. *Oncale v. Sundowner Offshore Services, Inc.,* 118 S. Ct. 998, 1002 (1998).

5. *Faragher,* 118 S. Ct. at 2283. However, when isolated incidents that are not "extremely serious" come to the attention of management, appropriate corrective action should still be taken so that they do not escalate. *See* Section V(C)(1)(a), below.

6. *Oncale,* 118 S. Ct. at 1003.

7. Some previous Commission documents classified harassment as either "quid pro quo" or hostile environment. However, it is now more useful to distinguish between harassment that results in a tangible employment action and harassment that creates a hostile work environment, since that dichotomy determines whether the employer can raise the affirmative defense to vicarious liability. Guidance on the definition of "tangible employment action" appears in section IV(B), below.

8. The guidance in this document applies to federal sector employers, as well as all other employers covered by the statutes enforced by the Commission.

9. 29 C.F.R. § 1604.11(d).

10. The Commission will rescind Subsection 1604.11(c) of the 1980 Guidelines on Sexual Harassment, 29 CFR § 1604.11(c). In addition, the following Commission guidance is no longer in effect: Subsection D of the 1990 Policy Statement on Current Issues in Sexual Harassment ("Employer Liability for Harassment by Supervisors"), EEOC Compliance Manual (BNA) N:4050–58 (3/19/90); and EEOC Compliance Manual Section 615.3(c) (BNA) 6:15-0007–0008.

The remaining portions of the 1980 Guidelines, the 1990 Policy Statement, and Section 615 of the Compliance Manual remain in effect. Other Commission guidance on harassment also remains in effect, including the Enforcement Guidance on *Harris v. Forklift Sys., Inc.,* EEOC Compliance Manual (BNA) N:4071 (3/8/94) and the Policy Guidance on Employer Liability for Sexual Favoritism, EEOC Compliance Manual (BNA) N:5051 (3/19/90).

11. Harassment that is targeted at an individual because of his or her sex violates Title VII even if it does not involve sexual comments or conduct. Thus, for example, frequent, derogatory remarks about women could constitute unlawful harassment even if the remarks are not sexual in nature. *See* 1990 Policy Guidance on Current Issues of Sexual Harassment, subsection C(4) ("sex-based harassment—that is, harassment not involving sexual activity or language—may also give rise to Title VII liability . . . if it is 'sufficiently patterned or pervasive' and directed at employees because of their sex").

12. "Protected activity" means opposition to discrimination or participation in proceedings covered by the anti-discrimination statutes. Harassment based on protected activity can constitute unlawful retaliation. *See* EEOC Compliance Manual Section 8 ("Retaliation") (BNA) 614:001 (May 20, 1998).

13. For cases applying *Ellerth* and *Faragher* to harassment on different bases, *see Hafford v. Seidner,* 167 F.3d 1074, 1080 (6th Cir. 1999) (religion and race); *Breeding v. Arthur J. Gallagher and Co.,* 164 F.3d 1151, 1158 (8th Cir. 1999) (age); *Allen v. Michigan Department of Corrections,* 165 F.3d 405, 411 (6th Cir. 1999) (race); *Richmond-Hopes v. City of Cleveland,* No. 97-3595, 1998 WL 808222 at *9 (6th Cir. Nov. 16, 1998) (unpublished) (retaliation); *Wright-Simmons v. City of Oklahoma City,* 155 F.3d 1264, 1270 (10th Cir. 1998) *(race); Gotfryd v. Book Covers, Inc.,* No. 97 C 7696, 1999 WL 20925 at *5 (N.D. Ill. Jan. 7, 1999) (national origin). *See also Wallin v. Minnesota Department of Corrections,* 153 F.3d 681, 687 (8th Cir. 1998) (assuming without deciding that ADA hostile environment claims are modeled after Title VII claims), *cert. denied,* 119 S. Ct. 1141 (1999).

14. The majority's analysis in both *Faragher* and *Ellerth* drew upon the liability standards for harassment on other protected bases. It is therefore clear that the same standards apply. *See Faragher,* 118 S. Ct. at 2283 (in determining appropriate standard of liability for sexual harassment by supervisors, Court "drew upon cases recognizing liability for discriminatory harassment based on race and national origin"); *Ellerth,* 118 S. Ct. at 2268 (Court imported concept of "tangible employment action" in race, age and national origin discrimination cases for resolution of vicarious liability in sexual harassment cases). *See also* cases cited in n.13, above.

15. *Ellerth,* 118 S. Ct. at 2270; *Faragher,* 118 S. Ct. at 2293.

16. Numerous statutes contain the word "supervisor," and some contain definitions of the term. *See, e.g.,* 12 U.S.C. § 1813(r) (definition of "State bank super-

visor" in legislation regarding Federal Deposit Insurance Corporation); 29 U.S.C. § 152(11) (definition of "supervisor" in National Labor Relations Act); 42 U.S.C. § 8262(2) (definition of "facility energy supervisor" in Federal Energy Initiative legislation). The definitions vary depending on the purpose and structure of each statute. The definition of the word "supervisor" under other statutes does not control, and is not affected by, the meaning of that term under the employment discrimination statutes.

17. *See* 42 U.S.C. 2000e(a) (Title VII); 29 U.S.C. 630(b) (ADEA); and 42 U.S.C. § 12111(5)(A) (ADA) (all defining "employer" as including any agent of the employer).

18. *Meritor Savings Bank, FSB v. Vinson*, 477 U.S. 57, 72 (1986); *Faragher*, 118 S. Ct. at 2290 n.3; *Ellerth*, 118 S. Ct. at 2266.

19. *See Faragher*, 118 S. Ct. at 2288 (analysis of vicarious liability "calls not for a mechanical application of indefinite and malleable factors set forth in the Restatement . . . but rather an inquiry into the reasons that would support a conclusion that harassing behavior ought to be held within the scope of a supervisor's employment . . . ") and at 2290 n.3 (agency concepts must be adapted to the practical objectives of the anti-discrimination statutes).

20. *Faragher*, 118 S. Ct. at 2290; *Ellerth*, 118 S. Ct. at 2269.

21. *Ellerth*, 118 S. Ct. at 2269.

22. *Ellerth*, 118 S. Ct. at 2269.

23. *Faragher*, 118 S. Ct. at 2280. For a more detailed discussion of the harassers' job responsibilities, *see Faragher*, 864 F. Supp. 1552, 1563 (S.D. Fla. 1994).

24. *See Grozdanich v. Leisure Hills Health Center*, 25 F. Supp.2d 953, 973 (D. Minn. 1998) ("it is evident that the Supreme Court views the term 'supervisor' as more expansive than as merely including those employees whose opinions are dispositive on hiring, firing, and promotion"; thus, "charge nurse" who had authority to control plaintiff's daily activities and recommend discipline qualified as "supervisor" and therefore rendered employer vicariously liable under Title VII for his harassment of plaintiff, subject to affirmative defense).

25. *See Ellerth*, 118 S. Ct. at 2268 ("If, in the unusual case, it is alleged there is a false impression that the actor was a supervisor, when he in fact was not, the victim's mistaken conclusion must be a reasonable one."); *Llampallas v. Mini-Circuit Lab, Inc.*, 163 F.3d 1236, 1247 (11th Cir. 1998) ("Although the employer may argue that the employee had no actual authority to take the employment action against the plaintiff, apparent authority serves just as well to impute liability to the employer for the employee's action.").

26. Of course, traditional principles of mitigation of damages apply in these cases, as well as all other employment discrimination cases. *See generally Ford Motor Co. v. EEOC*, 458 U.S. 219 (1982).

27. *Ellerth*, 118 S. Ct. at 2269; *Faragher*, 118 S. Ct. 2284–85. *See also Durham Life Insurance Co., v. Evans*, 166 F.3d 139, 152 (3rd Cir. 1999) ("A supervisor can only take a tangible adverse employment action because of the authority delegated by the employer . . . and thus the employer is properly charged with the consequences of that delegation.").

28. *Ellerth*, 118 S. Ct. at 2268.

29. All listed criteria are set forth in *Ellerth*, 118 S. Ct. at 2269.

30. All listed examples are set forth in *Ellerth* and/or *Faragher*. *See Ellerth*, 118 S. Ct. at 2268 and 2270; *Faragher*, 118 S. Ct. at 2284, 2291, and 2293.

31. Other forms of formal discipline would qualify as well, such as suspension. Any disciplinary action undertaken as part of a program of progressive discipline is "tangible" because it brings the employee one step closer to discharge.

32. The Commission disagrees with the Fourth Circuit's conclusion in *Reinhold v. Commonwealth of Virginia*, 151 F.3d 172 (4th Cir. 1998), that the plaintiff was not subjected to a tangible employment action where the harassing supervisor "dramatically increased her workload," *Reinhold*, 947 F. Supp. 919, 923 (E.D Va. 1996), denied her the opportunity to attend a professional conference, required her to monitor and discipline a co-worker, and generally gave her undesirable assignments. The Fourth Circuit ruled that the plaintiff had not been subjected to a tangible employment action because she had not "experienced a change in her employment status akin to a demotion or a reassignment entailing significantly different job responsibilities." 151 F.3d at 175. It is the Commission's view that the Fourth Circuit misconstrued *Faragher* and *Ellerth*. While minor changes in work assignments would not rise to the level of tangible job harm, the actions of the supervisor in *Reinhold* were substantial enough to significantly alter the plaintiff's employment status.

33. *See Durham*, 166 F.3d at 152–53 (assigning insurance salesperson heavy load of inactive policies, which had a severe negative impact on her earnings, and depriving her of her private office and secretary, were tangible employment actions); *Bryson v. Chicago State University*, 96 F.3d 912, 917 (7th Cir. 1996) ("Depriving someone of the building blocks for . . . a promotion . . . is just as serious as depriving her of the job itself.").

34. *See Flaherty v. Gas Research Institute*, 31 F.3d 451, 457 (7th Cir. 1994) (change in reporting relationship requiring plaintiff to report to former subordinate, while maybe bruising plaintiff's ego, did not affect his salary, benefits, and level of responsibility and therefore could not be challenged in ADEA claim), *cited in Ellerth*, 118 S. Ct. at 2269.

35. *See Crady v. Liberty Nat. Bank & Trust Co. of Ind.*, 993 F.2d 132, 136 (7th Cir. 1993) ("A materially adverse change might be indicated by a termination of employment, a demotion evidenced by a decrease in wage or salary, a less distinguished title, a material loss of benefits, significantly diminished material responsibilities, or other indices that might be unique to the particular situation."), *quoted in Ellerth*, 118 S. Ct. at 2268–69.

36. *See Nichols v. Frank*, 42 F.3d 503, 512–13 (9th Cir. 1994) (employer vicariously liable where its supervisor granted plaintiff's leave requests based on her submission to sexual conduct), *cited in Faragher*, 118 S. Ct. at 2285.

37. *See Ellerth*, 118 S. Ct. at 2268 and *Faragher*, 118 S. Ct. at 2284 (listed examples of tangible employment actions that included both positive and negative job decisions: hiring *and* firing; promotion *and* failure to promote).

38. The link could be established even if the harasser was not the ultimate decision maker. *See, e.g., Shager v Upjohn Co.*, 913 F.2d 398, 405 (7th Cir. 1990) (noting that committee rather than the supervisor fired plaintiff, but employer was still liable because committee functioned as supervisor's "cat's paw"), *cited in Ellerth*, 118 S. Ct. at 2269.

39. *Llampallas*, 163 F.3d at 1247.

40. *Ellerth*, 118 S. Ct. at 2270 ("[n]o affirmative defense is available . . . when the supervisor's harassment culminates in a tangible employment action . . ."); *Faragher*, 118 S. Ct. at 2293 (same). *See also Durham*, 166 F.3d at 154 ("When harassment

becomes adverse employment action, the employer loses the affirmative defense, even if it might have been available before."); *Lissau v. Southern Food Services, Inc.*, 159 F.3d 177, 184 (4th Cir. 1998) (the affirmative defense "is not available in a hostile work environment case when the supervisor takes a tangible employment action against the employee as part of the harassment") (Michael, J., concurring).

41. *Ellerth*, 118 S. Ct. at 2265. Even if the preceding acts were not severe or pervasive, they still may be relevant evidence in determining whether the tangible employment action was discriminatory.

42. *See Lissau v. Southern Food Service, Inc.*, 159 F.3d at 182 (if plaintiff could not prove that her discharge resulted from her refusal to submit to her supervisor's sexual harassment, then the defendant could advance the affirmative defense); *Newton v. Caldwell Laboratories*, 156 F.3d 880, 883 (8th Cir. 1998) (plaintiff failed to prove that her rejection of her supervisor's sexual advances was the reason that her request for a transfer was denied and that she was discharged; her claim was therefore categorized as one of hostile environment harassment); *Fierro v. Saks Fifth Avenue*, 13 F. Supp.2d 481, 491 (S.D.N.Y. 1998) (plaintiff claimed that his discharge resulted from national origin harassment but court found that he was discharged because of embezzlement; thus, employer could raise affirmative defense as to the harassment preceding the discharge).

43. *See Faragher*, 118 S. Ct. at 2292 ("If the victim could have avoided harm, no liability should be found against the employer who had taken reasonable care.").

44. *See, e.g., EEOC v. SBS Transit, Inc.*, No. 97-4164, 1998 WL 903833 at *1 (6th Cir. Dec. 18, 1998) (unpublished) (lower court erred when it reasoned that employer liability for sexual harassment is negated if the employer responds adequately and effectively once it has notice of the supervisor's harassment; that standard conflicts with affirmative defense which requires proof that employer "took reasonable care to *prevent* and correct promptly any sexually harassing behavior and that the plaintiff employee unreasonably failed to take advantage of preventative or corrective opportunities provided by the employer").

45. *Ellerth*, 118 S. Ct. at 2270.

46. *See Indest v. Freeman Decorating, Inc.*, 168 F.3d 795, 803 (5th Cir. 1999) ("when an employer satisfies the first element of the Supreme Court's affirmative defense, it will likely forestall its own vicarious liability for a supervisor's discriminatory conduct by nipping such behavior in the bud") (Wiener, J., concurring in *Indest*, 164 F.3d 258 (5th Cir. 1999)). The Commission agrees with Judge Wiener's concurrence in *Indest* that the court in that case dismissed the plaintiff's claims on an erroneous basis. The plaintiff alleged that her supervisor made five crude sexual comments or gestures to her during a week-long convention. She reported the incidents to appropriate management officials who investigated the matter and meted out appropriate discipline. No further incidents of harassment occurred. The court noted that it was "difficult to conclude" that the conduct to which the plaintiff was briefly subjected created an unlawful hostile environment. Nevertheless, the court went on to consider liability. It stated that *Ellerth* and *Faragher* do not apply where the plaintiff quickly resorted to the employer's grievance procedure and the employer took prompt remedial action. In such a case, according to the court, the employer's quick response exempts it from liability. The Commission agrees with Judge Wiener that *Ellerth* and *Faragher* do control the analysis in such cases, and that an employee's prompt complaint to management forecloses the employer from proving the affirmative defense. However, as Judge Wiener pointed out, an employer's quick remedial action will

often thwart the creation of an unlawful hostile environment, rendering any consideration of employer liability unnecessary.

47. *See Greene v. Dalton,* 164 F.3d 671, 674 (D.C. Cir. 1999) (in order for defendant to avoid all liability for sexual harassment leading to rape of plaintiff "it must show not merely that [the plaintiff] inexcusably delayed reporting the alleged rape . . . but that, as a matter of law, a reasonable person in [her] place would have come forward early enough to prevent [the] harassment from becoming 'severe or pervasive'").

48. *Ellerth,* 118 S. Ct. at 2267.

49. Under this same principle, it is the Commission's position that an employer is liable for punitive damages if its supervisor commits unlawful harassment or other discriminatory conduct with malice or with reckless indifference to the employee's federally protected rights. (The Supreme Court will determine the standard for awarding punitive damages in *Kolstad v. American Dental Association,* 119 S. Ct. 401 (1998) (granting certiorari).) The test for imposition of punitive damages is the mental state of the harasser, not of higher-level officials. This approach furthers the remedial and deterrent objectives of the anti-discrimination statutes, and is consistent with the vicarious liability standard set forth in *Faragher* and *Ellerth.*

50. Even if higher management proves that evidence it discovered after-the-fact would have justified the supervisor's action, such evidence can only limit remedies, not eliminate liability. *McKennon v. Nashville Banner Publishing Co.,* 513 U.S. 352, 360–62 (1995).

51. *See Faragher,* 118 S. Ct. at 2293, and *Ellerth,* 118 S. Ct. at 2270 (affirmative defense operates either to eliminate liability or limit damages).

52. *See Faragher,* 118 S. Ct. at 2292 ("if damages could reasonably have been mitigated no award against a liable employer should reward a plaintiff for what her own efforts could have avoided").

53. *See* Section V(C)(3) for a discussion of preventive and corrective care by small employers.

54. *See Hurley v. Atlantic City Police Dept.,* No. 96-5634, 96-5633, 96-5661, 96-5738, 1999 WL 150301 (3d Cir. March 18, 1999) ("*Ellerth* and *Faragher* do not, as the defendants seem to assume, focus mechanically on the formal existence of a sexual harassment policy, allowing an absolute defense to a hostile work environment claim whenever the employer can point to an anti-harassment policy of some sort"; defendant failed to prove affirmative defense where it issued written policies without enforcing them, painted over offensive graffiti every few months only to see it go up again in minutes, and failed to investigate sexual harassment as it investigated and punished other forms of misconduct.).

55. *See Dees v. Johnson Controls World Services, Inc.,* 168 F.3d 417, 422 (11th Cir. 1999) (employer can be held liable despite its immediate and appropriate corrective action in response to harassment complaint if it had knowledge of the harassment prior to the complaint and took no corrective action).

56. *Ellerth,* 118 S. Ct. at 2270.

57. A union grievance and arbitration system does not fulfill this obligation. Decision making under such a system addresses the collective interests of bargaining unit members, while decision making under an internal harassment complaint process should focus on the individual complainant's rights under the employer's anti-harassment policy.

An arbitration, mediation, or other alternative dispute resolution process also does not fulfill the employer's duty of due care. The employer cannot discharge its responsibility to investigate complaints of harassment and undertake corrective measures by providing employees with a dispute resolution process. For further discussion of the impact of such procedures on the affirmative defense, see Section V(D)(1)(b), below.

Finally, a federal agency's formal, internal EEO complaint process does not, by itself, fulfill its obligation to exercise reasonable care. That process only addresses complaints of violations of the federal EEO laws, while the Court, in *Ellerth*, made clear that an employer should encourage employees "to report harassing conduct before it becomes severe or pervasive." *Ellerth*, 118 S. Ct. at 2270. Furthermore, the EEO process is designed to assess whether the agency is liable for unlawful discrimination and does not necessarily fulfill the agency's obligation to undertake immediate and appropriate corrective action.

58. Although the affirmative defense does not apply in cases of harassment by co-workers or non-employees, an employer cannot claim lack of knowledge as a defense to such harassment if it did not make clear to employees that they can bring such misconduct to the attention of management and that such complaints will be addressed. *See Perry v. Ethan Allen*, 115 F.3d 143, 149 (2d Cir. 1997) ("When harassment is perpetrated by the plaintiff's coworkers, an employer will be liable if the plaintiff demonstrates that 'the employer either provided no reasonable avenue for complaint or knew of the harassment but did nothing about it'"), *cited in Faragher*, 118 S. Ct. at 2289. Furthermore, an employer is liable for harassment by a co-worker or non-employer if management knew or should have known of the misconduct, unless the employer can show that it took immediate and appropriate corrective action. 29 C.F.R. § 1604.11(d). Therefore, the employer should have a mechanism for investigating such allegations and undertaking corrective action, where appropriate.

59. Surveys have shown that a common reason for failure to report harassment to management is fear of retaliation. *See, e.g.,* Louise F. Fitzgerald & Suzanne Swan, "Why Didn't She Just Report Him? The Psychological and Legal Implications of Women's Responses to Sexual Harassment," 51 *Journal of Social Issues* 117, 121–22 (1995) (citing studies). Surveys also have shown that a significant proportion of harassment victims are worse off after complaining. *Id.* at 123–24; *see also* Patricia A. Frazier, "Overview of Sexual Harassment From the Behavioral Science Perspective," paper presented at the American Bar Association National Institute on Sexual Harassment at B-17 (1998) (reviewing studies that show frequency of retaliation after victims confront their harasser or filed formal complaints).

60. *See Wilson v. Tulsa Junior College*, 164 F.3d 534, 541 (10th Cir. 1998) (complaint process deficient where it permitted employees to bypass the harassing supervisor by complaining to director of personnel services, but the director was inaccessible due to hours of duty and location in separate facility).

61. *Faragher*, 118 S. Ct. at 2293 (in holding as matter of law that City did not exercise reasonable care to prevent the supervisors' harassment, Court took note of fact that City's policy "did not include any assurance that the harassing supervisors could be bypassed in registering complaints"); *Meritor Savings Bank, FSB v. Vinson*, 471 U.S. 57, 72 (1986).

62. *See* Wilson, 164 F.3d at 541 (complaint procedure deficient because it only required supervisors to report "formal" as opposed to "informal" complaints of harassment); *Varner v. National Super Markets Inc.*, 94 F.3d 1209, 1213 (8th Cir. 1996), *cert. denied*, 519 U.S. 1110 (1997) (complaint procedure is not effective if it

does not require supervisor with knowledge of harassment to report the information to those in position to take appropriate action).

63. It is particularly important for federal agencies to explain the statute of limitations for filing formal EEO complaints, because the regulatory deadline is only 45 days and employees may otherwise assume they can wait whatever length of time it takes for management to complete its internal investigation.

64. If an employer actively misleads an employee into missing the deadline for filing a charge by dragging out its investigation and assuring the employee that the harassment will be rectified, then the employer would be "equitably estopped" from challenging the delay. *See Currier v. Radio Free Europe/Radio Liberty, Inc.,* 159 F.3d 1363, 1368 (D.C. Cir. 1998) ("an employer's affirmatively misleading statements that a grievance will be resolved in the employee's favor can establish an equitable estoppel"); *Miranda v. B & B Cash Grocery Store, Inc.,* 975 F.2d 1518, 1531 (11th Cir. 1992) (tolling is appropriate where plaintiff was led by defendant to believe that the discriminatory treatment would be rectified); *Miller v. Beneficial Management Corp.,* 977 F.2d 834, 845 (3d Cir. 1992) (equitable tolling applies where employer's own acts or omission has lulled the plaintiff into foregoing prompt attempt to vindicate his rights).

65. The sharing of records about a harassment complaint with prospective employers of the complainant could constitute unlawful retaliation. *See* Compliance Manual Section 8 ("Retaliation), subsection II D (2), (BNA) 614:0005 (5/20/98).

66. One court has suggested that it may be permissible to honor such a request, but that when the harassment is severe, an employer cannot just stand by, even if requested to do so. *Torres v. Pisano,* 116 F.3d 625 (2d Cir.), *cert. denied,* 118 S. Ct. 563 (1997).

67. Employers may hesitate to set up such a phone line due to concern that it may create a duty to investigate anonymous complaints, even if based on mere rumor. To avoid any confusion as to whether an anonymous complaint through such a phone line triggers an investigation, the employer should make clear that the person who takes the calls is not a management official and can only answer questions and provide information. An investigation will proceed only if a complaint is made through the internal complaint process or if management otherwise learns about alleged harassment.

68. *See, e.g., Van Zant v. KLM Royal Dutch Airlines,* 80 F.3d 708, 715 (2d Cir. 1996) (employer's response prompt where it began investigation on the day that complaint was made, conducted interviews within two days, and fired the harasser within ten days); *Steiner v. Showboat Operating Co.,* 25 F.3d 1459, 1464 (9th Cir. 1994) (employer's response to complaints inadequate despite eventual discharge of harasser where it did not seriously investigate or strongly reprimand supervisor until after plaintiff filed charge with state FEP agency), *cert. denied,* 513 U.S. 1082 (1995); *Saxton v. AT&T,* 10 F.3d 526, 535 (7th Cir 1993) (investigation prompt where it was begun one day after complaint and a detailed report was completed two weeks later); *Nash v. Electrospace Systems, Inc.* 9 F.3d 401, 404 (5th Cir. 1993) (prompt investigation completed within one week); *Juarez v. Ameritech Mobile Communications, Inc.,* 957 F.2d 317, 319 (7th Cir. 1992) (adequate investigation completed within four days).

69. Management may be reluctant to release information about specific disciplinary measures that it undertakes against the harasser, due to concerns about potential defamation claims by the harasser. However, many courts have recognized that limited disclosures of such information are privileged. For cases addressing defenses

to defamation claims arising out of alleged harassment, *see Duffy v. Leading Edge Products,* 44 F.3d 308, 311 (5th Cir. 1995) (qualified privilege applied to statements accusing plaintiff of harassment); *Garziano v. E.I. DuPont de Nemours & Co.,* 818 F.2d 380 (5th Cir. 1987) (qualified privilege protects employer's statements in bulletin to employees concerning dismissal of alleged harasser); *Stockley v. AT&T,* 687 F. Supp. 764 (E.D.N.Y. 1988) (statements made in course of investigation into sexual harassment charges protected by qualified privilege).

70. *Mockler v Multnomah County,* 140 F.3d 808, 813 (9th Cir. 1998).

71. In some cases, accused harassers who were subjected to discipline and subsequently exonerated have claimed that the disciplinary action was discriminatory. No discrimination will be found if the employer had a good faith belief that such action was warranted and there is no evidence that it undertook less punitive measures against similarly situated employees outside his or her protected class who were accused of harassment. In such circumstances, the Commission will not find pretext based solely on an after-the-fact conclusion that the disciplinary action was inappropriate. *See Waggoner v. City of Garland Tex.,* 987 F.2d 1160, 1165 (5th Cir. 1993) (where accused harasser claims that disciplinary action was discriminatory, "[t]he real issue is whether the employer reasonably believed the employee's allegation [of harassment] and acted on it in good faith, or to the contrary, the employer did not actually believe the co-employee's allegation but instead used it as a pretext for an otherwise discriminatory dismissal").

72. *See Steiner v. Showboat Operating Co.,* 25 F.3d 1459, 1464 (9th Cir. 1994) (employer remedial action for sexual harassment by supervisor inadequate where it twice changed plaintiff's shift to get her away from supervisor rather than change his shift or work area), *cert. denied,* 513 U.S. 1082 (1995).

73. *See Guess v. Bethlehem Steel Corp.,* 913 F.2d 463, 465 (7th Cir. 1990) ("a remedial measure that makes the victim of sexual harassment worse off is ineffective *per se*").

74. An oral warning or reprimand would be appropriate only if the misconduct was isolated and minor. If an employer relies on oral warnings or reprimands to correct harassment, it will have difficulty proving that it exercised reasonable care to prevent and correct such misconduct.

75. *See Varner,* 94 F.3d at 1213 (complaint procedure is not effective if it does not require supervisor with knowledge of harassment to report the information to those in position to take appropriate action), *cert. denied,* 117 S. Ct. 946 (1997); *accord Wilson v. Tulsa Junior College,* 164 F.3d at 541.

76. *See Wilson,* 164 F.3d at 541 (complaint procedure deficient because it only required supervisors to report "formal" as opposed to "informal" complaints of harassment).

77. *See, e.g., Splunge v. Shoney's, Inc.,* 97 F.3d 488, 490 (11th Cir. 1996) (where harassment of plaintiffs was so pervasive that higher management could be deemed to have constructive knowledge of it, employer was obligated to undertake corrective action even though plaintiffs did not register complaints); *Fall v. Indiana Univ. Bd. of Trustees,* 12 F. Supp.2d 870, 882 (N.D. Ind. 1998) (employer has constructive knowledge of harassment by supervisors where it "was so broad in scope and so permeated the workplace that it must have come to the attention of someone authorized to do something about it").

78. In *Faragher,* the City lost the opportunity to establish the affirmative defense in part because "its officials made no attempt to keep track of the conduct of supervisors." *Faragher,* 118 S. Ct. at 2293.

79. See subsections V(C)(1)(e)(ii) and V(C)(2), above.

80. If the owner of the business commits unlawful harassment, then the business will automatically be found liable under the alter ego standard and no affirmative defense can be raised. *See* Section VI, below.

81. *Faragher,* 118 S. Ct. at 2293.

82. *Faragher,* 118 S. Ct. at 2292 ("If the victim could have avoided harm, no liability should be found against the employer who had taken reasonable care, and if damages could reasonably have been mitigated no award against a liable employer should reward a plaintiff for what her own efforts could have avoided.").

83. *Ellerth,* 118 S. Ct. at 2270; *Faragher,* 118 S. Ct. at 2293. *See also Scrivner v. Socorro Independent School District,* 169 F.3d 969, 971 (5th Cir., 1999) (employer established second prong of defense where harassment began during summer, plaintiff misled investigators inquiring into anonymous complaint by denying that harassment occurred, and plaintiff did not complain about the harassment until the following March).

84. The employee is not required to have chosen "the course that events later show to have been the best." Restatement (Second) of Torts § 918, comment c.

85. *See Corcoran v. Shoney's Colonial, Inc.,* 24 F. Supp.2d 601, 606 (W.D. Va. 1998) ("Though unwanted sexual remarks have no place in the work environment, it is far from uncommon for those subjected to such remarks to ignore them when they are first made.").

86. *See Faragher,* 118 S. Ct. at 2292 (defense established if plaintiff unreasonably failed to avail herself of "a proven, effective mechanism for reporting and resolving complaints of sexual harassment, available to the employee without undue risk or expense"). *See also* Restatement (Second) of Torts § 918, comment c (tort victim "is not barred from full recovery by the fact that it would have been reasonable for him to make expenditures or subject himself to pain or risk; it is only when he is unreasonable in refusing or failing to take action to prevent further loss that his damages are curtailed").

87. *See* n. 59, above.

88. *See Faragher,* 118 S. Ct. at 2292 (employee should not recover for harm that could have been avoided by utilizing a proven, effective complaint process that was available "without undue risk or expense").

89. *See Wilson,* 164 F.3d at 541 (complaint process deficient where official who could take complaint was inaccessible due to hours of duty and location in separate facility).

90. *See* Policy Statement on Mandatory Binding Arbitration of Employment Discrimination Disputes as a Condition of Employment, EEOC Compliance Manual (BNA) N:3101 (7/10/97).

91. For a discussion of defamation claims and the application of a qualified privilege to an employer's statements about instances of harassment, *see* n. 69, above.

92. *See Watts v. Kroger Company,* 170 F.3d 505, 510 (5th Cir., 1999) (plaintiff made effort "to avoid harm otherwise" where she filed a union grievance and did not utilize the employer's harassment complaint process; both the employer and union procedures were corrective mechanisms designed to avoid harm).

93. Both the staffing firm and the client may be legally responsible, under the anti-discrimination statutes, for undertaking corrective action. *See* Enforcement Guidance: Application of EEO Laws to Contingent Workers Placed by Temporary Employment Agencies and Other Staffing Firms, EEOC Compliance Manual (BNA) N:3317 (12/3/97).

94. *See also Ellerth*, 118 S. Ct. at 2267 (under agency principles an employer is indirectly liable "where the agent's high rank in the company makes him or her the employer's alter ego"); *Harrison v. Eddy Potash, Inc.*, 158 F.3d 1371, 1376 (10th Cir. 1998) ("the Supreme Court in Burlington acknowledged an employer can be held vicariously liable under Title VII if the harassing employee's 'high rank in the company makes him or her the employer's alter ego'").

95. *Faragher*, 118 S. Ct. at 2284.

96. The Court noted that the standards for employer liability were not at issue in the case of *Harris v. Forklift Systems*, 510 U.S. 17 (1993), because the harasser was the president of the company. *Faragher*, 118 S. Ct. at 2284.

97. An individual who has an ownership interest in an organization, receives compensation based on its profits, and participates in managing the organization would qualify as an "owner" or "partner." *Serapion v. Martinez*, 119 F.3d 982, 990 (1st Cir. 1997), *cert. denied*, 118 S. Ct. 690 (1998).

98. *Id.*

APPENDIX E

Sample EEOC Deposition of Defendant[1]

IN THE UNITED STATES DISTRICT COURT
FOR THE NORTHERN DISTRICT _____
EASTERN DIVISION

EQUAL EMPLOYMENT OPPORTUNITY)
COMMISSION,)
)
)
Plaintiff,) CIVIL ACTION NO. _____
)
v.)
) Judge _____
_____ CORPORATION,) Magistrate Judge _____
)
Defendant.)
_____)

Pursuant to Rule 30(b)(6) of the Federal Rules of Civil Procedure, notice is hereby given the Plaintiff, United States Equal Employment Opportunity Commission, will take the deposition upon oral examination of Defendant at the offices of the undersigned counsel for the EEOC, _____ on September _____ . The deposition will begin at 9:00 a.m., and shall continue from the date listed above until completed.

In accordance with Fed. R. Civ. P. 30(b)(6), Defendant shall designate one or more officials, officers, managing agents or other persons to testify on their behalf and, for each such individual, Defendant shall designate the matters about which the individual shall testify.

Deposition Subject Areas

The deposition examination will be on the following subjects and except where otherwise stated, shall cover the period from January 1, 1996, to the present:

1. Defendant's practices, procedures, or policies concerning the following subjects:
   a. job promotions;
   b. hiring and salaries for _____ managers;
   c. discipline;
   d. terminations;
   e. any language policies;
   f. the assignment of work stations;
   g. the assignment of clients;
   h. the personal use of _____ products or services by employees;
   i. performance of free services by employees;
   j. client development (e.g., promotional sales and client discounts);
   k. ordering products for _____ ;
   l. _____ _____ _____ _____ _____ _____ ;
   m. the maintenance of personnel records.
2. Training provided to Defendant's employees concerning the subjects listed in topic number 3, above.
3. The manner in which Defendant determined the commission rate and pay rate for _____ .
4. _____ _____ _____ _____ _____ _____ .
5. Explain the following documents produced by Defendant:
   a. _____ _____ ;
   b. the employee time recaps; and
   c. ticket details.
6. Measures taken by Defendant to ensure _____ and/or Defendant as a whole is in compliance with anti-discrimination laws including, but not limited to, the following subjects:
   a. the existence, organization, duties, responsibilities, and functions of any department(s), division(s), office(s) or employee(s) of Defendant charged with ensuring compliance with anti-discrimination laws;
   b. criteria used to select Defendant's personnel responsible for ensuring compliance with anti-discrimination laws;
   c. the identities, qualifications, and training of individuals responsible for ensuring compliance with anti-discrimination laws; and
   d. resources (staff availability and monetary) available and/or allocated by Defendant to investigate allegations of discrimination.
   e. resources (staff availability and monetary) available and/or allocated by Defendant to provide training on anti-discrimination laws;
   f. Defendant's anti-discrimination policy and its method(s) for informing employees of the policy;

    g. Defendant's non-discrimination statement _____ and

    h. for the time period from January 1, 1991 to the present, all training provided to employees _____ and all individuals with any managerial responsibility for _____ concerning discrimination against employees or customers.

7. The practice(s), policy(ies), or procedure(s) that employees are supposed to follow to file complaints of discrimination, and Defendant's method(s) for informing employees of the practice(s), policy(ies), or procedure(s).

8. Defendant's practice(s), policy(ies), or procedure(s) for investigating complaints of discrimination, and Defendant's method(s) for informing employees of the practice(s), policy(ies), or procedure(s).

9. Actions, if any, taken by Defendant in response to complaints of discrimination filed with either the _____ on Human Rights or the Equal Employment Opportunity Commission.

10. Changes to any practice(s), policy(ies), or procedure(s) of Defendant made, formally or informally, in connection with or as a result of any law suit, charge, or internal complaint identified in response to Interrogatory Number 11 of Plaintiff EEOC's Second Set of Interrogatories.

Note

1. This amended notice reflects the re-scheduled date and time on which this deposition will occur and adds two subject areas, 1m and 11, to the original notice.

APPENDIX F

Connecticut Regulations Regarding Sexual Harassment Posting and Training Requirements

Regulations

Sections 46a-54-200 through 207, inclusive

Regulations provided below are for informational purposes ONLY. For official citations please refer to the **Regulations of Connecticut State Agencies.**

Section 46a-54-200. DEFINITIONS

For purposes of sections 46a-54-200 through 46a-54-207, inclusive:

(a) "Sexual Harassment" means any unwelcome sexual advances or requests for sexual favors or any conduct of a sexual nature when 1) submission to such conduct is made either explicitly or implicitly a term or condition of an individual's employment, 2) submission to or rejection of such conduct is by an individual is used as the basis for employment decisions affecting such individual, or 3) such conduct has the purpose or effect of substantially interfering with the individual's work performance or creating an intimidating, hostile or offensive working environment.

(b) "Employer" includes the state and all political subdivisions thereof, including the General Assembly, and means any person or employer with three or more persons in his employ.

(c) "Employer Having Fifty or More Employees" means the state and all political subdivisions thereof, including the General Assembly, and means any person or employer who has a total of fifty or more persons, including supervisory and managerial employees and partners, in his employ for a minimum of thirteen weeks during the training year.

(d) "Employee" means any person employed by an employer, but shall not include any individual employed by his parents, spouse or child, or in the domestic service of any person.

(e) "Supervisory Employee" means any individual who has the authority, by using her or his independent judgment, in the interest of the employer, to hire, suspend, lay off, promote, discharge, assign, nor discipline other employees, or responsibility to direct them, or to adjust their grievances or effectively to recommend such actions.

(f) "Commission" means the Commission on Human Rights and Opportunities created by section 46a-52 of the Connecticut General Statutes.

(g) "Training year" means the period of time from October first in any calendar year through September thirtieth in the following calendar year.

Section 46a-54-201. POSTING REQUIREMENT FOR EMPLOYERS HAVING THREE OR MORE EMPLOYEES

a. Employers with three or more employees must post notices to employees concerning the illegality of sexual harassment and remedies available to victims of sexual harassment.

b. Such information shall include, but is not limited to:
   1. The statutory definition of sexual harassment and examples of different types of sexual harassment
   2. Notice that sexual harassment is prohibited by the State of Connecticut's Discriminatory Employment Practices Law, subdivision (8) of subsection (a) of section 46a-60 of the Connecticut General Statutes;
   3. Notice that sexual harassment is prohibited by Title VII of the 1964 Civil Rights Act, as amended, 42 United States Code section 2000e *et. seq.*, and
   4. The remedies available, including but not limited to:
      A. Cease and desist orders,
      B. Back pay,
      C. Compensatory damages, and
      D. Hiring, promotion or reinstatement;
   5. Language to the effect that persons who commit sexual harassment may be subject to civil or criminal penalties;
   6. The address and telephone number of the Connecticut Commission on Human Rights and Opportunities; and
   7. A statement that Connecticut law requires that a formal written complaint be filed with the Commission within one hundred and eighty days of the date when the alleged sexual harassment occurred; and
   8. Any and all notices so posted will have the heading, **"SEXUAL HARASSMENT IS ILLEGAL"** in large bold-faced type.

c. The Commission strongly recommends, but does not require, that the poster include:

   1. A statement concerning the employer's policies and procedures regarding sexual harassment and a statement concerning the discipli-

nary action that may be taken if sexual harassment has been committed; and

2. A contact person at the place of employment to whom one can report complaints of sexual harassment or direct one's concerns regarding sexual harassment;

d. A model poster is appended to these regulations, labeled Appendix A.

Section 46a-54-202. WHERE TO POST

Employers must place, and keep notices in prominent and accessible locations upon its premises where notices to employees are customarily posted. Notices must be posted at each employer facility in such a manner that all employees and applicants at that facility will have the opportunity to see the notices on a regular basis.

Section 46a-54-203. WHEN TO POST

a. All employers with three or more employees shall post notices as soon as practicable after the effective date of these regulations, but no later than forty-five (45) days after the effective date of these regulations.

b. An employer shall promptly replace notices that are removed, destroyed or defaced.

Section 46a-54-204. POSTING AND TRAINING REQUIREMENTS FOR EMPLOYERS HAVING FIFTY OR MORE EMPLOYEES

a. An employer having fifty (50) or more employees shall comply with the posting requirements set forth in sections 46a-200 through 46a-54-207, inclusive.

b. An employer having fifty (50) or more employees must also provide two hours of training and education to all supervisory employees of employees in the State of Connecticut no later than October 1, 1993 and to all new supervisory employees of employees in the State of Connecticut within six months of their assumption of a supervisory position. Nothing in these regulations shall prohibit an employer from providing more than two hours of training and education.

c. Such training and education shall be concluded in a classroom-like setting, using clear and understandable language and in a format that allows participants to ask questions and receive answers. Audio, video and other teaching aids may be utilized to increase comprehension or to otherwise enhance the training process.

1. The content of the training shall include the following:

A. Describing the federal and state statutory provisions prohibiting sexual harassment in the work place with which the employer is required to comply, including, but not limited to, the Connecticut Discriminatory Employment Practices statute (section 46a-60 of the Connecticut General Statutes) and "Title VII of the Civil Rights Act of 1964, as amended (42 U.S.C. section 2000e, and following sections);

B. Defining sexual harassment as explicitly set forth in subdivision (8) of subsection (a) of section 46a-60 of the Connecticut General Statutes and as distinguished from other forms of illegal harassment prohibited by subsection (a) of section 46a-60 of the Connecticut General Statutes and section 3 of Public Act 91-58;

C. Discussing the types of conduct that may constitute sexual harassment under the law, including the fact that the harasser or the victim of harassment may be either a man or a woman and that harassment can occur involving persons of the same or opposite sex;

D. Describing the remedies available in sexual harassment cases, including, but not limited to, cease and desist orders; hiring, promotion or reinstatement; compensatory damages and back pay;

E. Advising employees that individuals who commit acts of sexual harassment may be subject to both civil and criminal penalties; and

F. Discussing strategies to prevent sexual harassment in the work place.

2. While not exclusive, the training may also include, but is not limited to, the following elements:

A. Informing training participants that all complaints of sexual harassment must be taken seriously, and that once a complaint is made, supervisory employees should report it immediately to officials designated by the employer, and that the contents of the complaint are personal and confidential and are not to be disclosed except to those persons with a need to know;

B. Coding experiential exercises such as role playing, coed group discussions and behavior modeling to facilitate understanding of what constitutes sexual harassment and how to prevent it;

C. Teaching the importance of interpersonal skills such as listening and bringing participants to understand what a person who is sexually harassed may be experiencing;

D. Advising employees of the importance of preventive strategies to avoid the negative effects sexual harassment has upon both the victim and the overall productivity of the work place due to interpersonal conflicts, poor absenteeism, turnover and grievances;

E. Explaining the benefits of learning about and eliminating sexual harassment which include a more positive work environment with greater productivity and potentially lower exposure to liability, in that employers—and supervisors personally—

have been held liable when it is shown that they knew or should have known of the harassment;

F. Explaining the employer's policy against sexual harassment, including a description of the procedures available for reporting instances of sexual harassment and the types of disciplinary actions which can and will be taken against persons who have been found to have engaged in sexual harassment; and

G. Discussing the perceptual and communication differences among all persons and, in this context, the concepts of "reasonable woman" and "reasonable man" developed in federal sexual harassment cases.

d. While not required by these regulations, the Commission encourages an employer having fifty (50) or more employees to provide an upgrade of legal interpretations and related developments concerning sexual harassment to supervisory personnel once every three (3) years.

46a-54-205. EFFECT OF PRIOR TRAINING

An employer is not required to train supervisory personnel who have received training after October 1, 1991 that:

1. substantially complies with the required content of the training set forth in subsection (c)(1) of section 46a 54-204; and
2. was provided in a classroom setting and lasted at least two hours.

Section 46a-54-206. TRAINERS

An employer required to provide training by these regulations may utilize individuals employed by the employer or other persons who agree to provide the required training, with or without reimbursement.

Section 46a-54-207. RECORD KEEPING

a. The Commission encourages each employer required to conduct training pursuant to Public Act 92-85 to maintain records concerning all training provided.

b. Such records; may include, but are not limited to:

1. documents sufficient to show the content of the training given, such as the curriculum;
2. the names, addresses and qualifications of the person conducting the training;
3. the names and titles of the personnel trained and the date or dates that each individual was trained;

c. The Commission encourages employers to maintain any such records for a minimum of one year, or if a discriminatory practice complaint is filed involving personnel trained, until such time as such complaint is finally resolved.

(Appendix A: model poster)

SEXUAL HARASSMENT
IS ILLEGAL

And is prohibited by the Connecticut Discriminatory Employment Practices Act (Section 46a-60(A)(8) of the Connecticut General Statutes) and Title VII of the Civil Rights Act of 1964 (42 United States Code Section 2000e *Et. Seal*).

Sexual Harassment means "any unwelcome sexual advances or requests for sexual favors or any conduct of a sexual nature when:

(1) Submission to such conduct is made either explicitly or implicitly a term or condition of an individual's employment.

(2) Submission to or rejection of such conduct by an individual is used as the basis for employment decisions affecting such individual; or

(3) Such conduct has the purpose or effect of substantially interfering with an individual's work performance or creating an intimidating, hostile or offensive working environment."

Examples of Sexual Harassment include: Unwelcome sexual advances; suggestive or lewd remarks; unwanted hugs, touches, kisses; requests for sexual favors; retaliation for complaining about sexual harassment; derogatory or pornographic posters, cartoons, drawings, or e-mail messages.

Remedies for Sexual harassment may include: cease and desist orders; back pay; compensatory damages; hiring, promotion or reinstatement. Individuals who engage in acts of sexual harassment may also be subject to civil and criminal penalties.

If you feel that you have been discriminated against, contact:

The Connecticut Commission on Human Rights and Opportunities (CHRO), 21 Grand Street, Hartford, Connecticut 06106. (Telephone Number 860-541-3400 or 800-477-5737). Connecticut law requires that a formal written complaint be filed with the commission within 180 days of the date when alleged harassment occurred.

(Optional)

Contact {Employer's Representative} if you have questions or concerns or believe that you or others are being sexually harassed.

Names   Telephone Number
Unit

If you need additional information contact:

The Permanent Commission on the Status of Women
18-20 Trinity Street
Hartford, Connecticut 06106

Telephone Number (860) 240-8300

e-mail: pcsw@po.state.ct.us

Website: www.cga.state.ct.us/pcsw/

APPENDIX G

Protected Classifications under State Fair Employment Practice Laws

Note: The following list includes only protected classifications under state fair employment practice laws. There may be other protected characteristics found in other state statutes.

Does not include protection for public employees, genetic testing prohibitions, and AIDS issues. Also, pregnancy is not cited as a protected classification, as it is generally covered under state sex discrimination prohibitions.

Alabama Has no comprehensive state fair employment practice law.	• Age (under separate statute)
Alaska	• Age • Race • Color • National Origin • Religion • Sex • Parenthood • Marital Status • Disability
Arizona	• Age • Race • Color • National Origin • Religion • Sex • Disability

Arkansas	• Race • National Origin • Religion • Gender • Physical Disability
California Requires employers to obtain sexual harassment information sheets from the Department of Fair Employment & Housing and distribute to employees. Program to eliminate sexual harassment required. Training is strongly encouraged.	• Race • Color • Religious Creed • National Origin • Sex • Marital Status • Age • Sexual Orientation • Ancestry • Disability • Medical Condition
Colorado	• Age • Creed • Race • Color • National Origin • Ancestry • Sex • Disability
Connecticut	• Age • Race • Color • National Origin • Present or past history of mental disorder • Mental Retardation • Ancestry • Religious Creed • Sex • Sexual Orientation • Marital Status • Physical Disability
Delaware	• Age • Race • Color • National Origin • Religion • Sex

	• Marital Status • Disability (under separate statute)
District of Columbia	• Race • Color • Religion • National Origin • Sex • Age • Marital Status • Personal Appearance • Sexual Orientation • Family Responsibilities • Disability • Matriculation • Political Affiliation • Source of Income • Place of Residence
Florida	• Age • Race • Color • National Origin • Religion • Sex • Marital Status • Disability
Georgia The Georgia Fair Employment Practice Law prohibits discrimination in public employment. However, Georgia has an equal pay act that applies to private employers and an age-discrimination statute that applies to all employers conducting business within the state.	• Age • Sex-based wage discrimination • Disability
Hawaii	• Age • Race • Color • Ancestry • Religion • Sex • Sexual Orientation • Marital Status

Hawaii, continued	• Disability • Arrest & Court Records
Idaho	• Age • Race • Color • National Origin • Religion • Sex • Disability
Illinois	• Age • Race • Citizenship Status • Color • National Origin • Ancestry • Religion • Sex • Marital Status • Military Status • Disability
Indiana	• Age (under separate statute) • Race • Color • National Origin • Ancestry • Religion • Sex • Disability
Iowa	• Age • Race • Color • Creed • National Origin • Religion • Sex • Disability
Kansas	• Race • Color • National Origin • Ancestry • Religion • Sex • Disability

	• Mental Disorder • Age (under separate statute)
Kentucky	• Age • Race • Color • National Origin • Religion • Sex • Familial Status • Disability • Smokers
Louisiana	• Age • Race • Color • National Origin • Religion • Sex • Disability • Sickle Cell Trait
Maine An employer with 15 or more employees must conduct sexual harassment training for all new employees within one year of employment.	• Age • Race • Color • National Origin • Ancestry • Religion • Sex • Disability
Maryland	• Age • Race • Color • National Origin • Religion • Sex • Marital Status • Disability
Massachusetts Every employer must adopt a sexual harassment policy. Employers are encouraged to conduct an education and training program for employees within one year of employment.	• Age • Race • Color • National Origin • Ancestry • Religious Creed • Sex • Sexual Orientation • Disability

Michigan	• Age • Race • Color • National Origin • Religion • Sex • Marital/Familial Status • Disability (under separate statute) • Height & Weight
Minnesota	• Age • Race • Color • National Origin • Religion • Creed • Sex • Sexual Orientation • Marital Status • Disability • Status with regard to public assistance
Mississippi Mississippi has no comprehensive state fair employment practice law.	
Missouri	• Age • Race • Color • National Origin • Ancestry • Religion • Sex • Disability
Montana	• Age • Race • Color • National Origin • Religion • Creed • Sex • Marital Status • Disability

Nebraska	• Race
	• Color
	• Age (under separate state statute)
	• National Origin
	• Religion
	• Sex
	• Marital Status
	• Disability
Nevada	• Age
	• Race
	• Sexual Orientation
	• Color
	• National Origin
	• Religion
	• Sex
	• Disability
New Hampshire	• Age
	• Race
	• Color
	• National Origin
	• Religious Creed
	• Sex
	• Sexual Orientation
	• Marital Status
	• Disability
New Jersey	• Age
	• Race
	• Color
	• National Origin
	• Ancestry
	• Creed
	• Sex
	• Sexual Orientation
	• Marital Status
	• Disability
	• Familial Status
	• Armed Forces
	• Atypical Hereditary Cellular or Blood Traits
New Mexico	• Age
	• Race
	• Color
	• National Origin

New Mexico, continued	• Ancestry • Religion • Sex • Marital Status • Disability
New York	• Age • Race • Color • National Origin • Creed • Sex • Marital Status • Disability
North Carolina	• Age • Race • Color • National Origin • Religion • Sex • Disability
North Dakota	• Age • Race • Color • National Origin • Religion • Sex • Marital Status • Disability • Public Assistance Status • Participation in lawful activities during nonworking hours
Ohio	• Age • Race • Color • National Origin • Ancestry • Religion • Sex • Disability
Oklahoma	• Age • Race • Color • National Origin

	• Religion • Sex • Disability
Oregon	• Age • Expunged Juvenile Record • Race • Color • National Origin • Religion • Sex • Marital Status • Disability • Family Relationships
Pennsylvania	• Age • Race • Color • National Origin • Ancestry • Religious Creed • Sex • Disability • Familial Status
Rhode Island Every employer must adopt a sexual harassment policy. Employers are encouraged to conduct an education and training program for employees within one year of employment.	• Age • Race • Color • Ancestral Origin • Religion • Sex • Sexual Orientation • Disability • Gender Identity
South Carolina	• Age • Race • Color • National Origin • Religion • Sex • Disability
South Dakota	• Race • Color • National Origin • Ancestry • Religion • Creed

| South Dakota, continued | • Sex |
| | • Disability |

Tennessee	• Age
	• Race
	• Color
	• National Origin
	• Creed
	• Religion
	• Sex
	• Disability (under separate statute)

Texas	• Age
	• Race
	• Color
	• National Origin
	• Religion
	• Sex
	• Disability

Utah	• Age
	• Race
	• Color
	• National Origin
	• Religion
	• Sex
	• Disability

Vermont Every employer must adopt a sexual harassment policy. Employers are encouraged to conduct an education and training program for employees within one year of employment.	• Age
	• Race
	• Color
	• National Origin
	• Ancestry
	• Religion
	• Sex
	• Sexual Orientation
	• Disability
	• Place of Birth

Virginia	• Age
	• Race
	• Color
	• National Origin
	• Religion
	• Sex
	• Marital Status
	• Disability

Washington	• Age • Race • Color • National Origin • Ancestry • Creed • Sex • Marital Status • Disability
West Virginia Encourages sexual harassment training	• Age • Race • Color • National Origin • Ancestry • Religion • Sex • Blindness • Disability
Wisconsin	• Age • Race • Color • National Origin • Ancestry • Religion • Sex • Creed • Sexual Orientation • Marital Status • Disability • Arrest or Conviction Record • Use or non-use of lawful product • Armed Forces
Wyoming	• Age • Race • Color • National Origin • Ancestry • Creed • Sex • Disability • Tobacco use

State laws are constantly evolving. This document should not be construed as legal advice or a legal opinion on any specific facts or circumstances, and is intended for general information purposes only. Please consult a lawyer when researching specific state laws, and protected classification under such laws.

APPENDIX H

Design Chart

Course Name: Harassment Prevention Training
Audience: Managers and Supervisors
Time: Three Hours (165 minutes of content)
Breaks: One/Fifteen Minutes

Content	Suggested Tools	Risk Activity Level	Estimated Time
Meet and Greet	*Music *Tent Name Card with Interactive Activity: House Rules: What was a rule in your house growing up? What is a rule in your house now? and Ground Rules	Low Risk Individual Activity	10 minutes
	*Pre-Test 10 True/False Questions with Communication Cards (To be debriefed at end of training)	Low to Medium Risk Individual/Group Activity	
Welcome and Introduction • To Trainer • To Agenda • To Ground Rules • To What You Will Get	*Flip Chart To debrief agenda *Tent Name Card To debrief Ground Rules	Low Risk Group Activity	15 minutes

(continued)

Content	Suggested Tools	Risk Activity Level	Estimated Time
Our Policy	*Music *Policy Scavenger Hunt (Six Questions) • Who is protected? • Give three (3) examples of harassing conduct based on sex • Give three (3) examples of harassing conduct based on race or national origin • Give three (3) examples of retaliation • Give three (3) examples of your responsibilities under the policy • Who can an employee complain to?	Low to Medium Risk Group Activity	10 minutes
Identifying Prohibited Conduct Quid Pro Quo **Teaching Points** • Definition of Quid Pro Quo Harassment • Sexual Conduct by Manager or Supervisor • Not Just About Sex • Unwelcome Threat to Job or Job-Related Benefit • For Managers and Supervisors: Subtle Quid Pro Quo Issues and Dating Relationships	**• Continuum or Conduct Line** *Flip Chart with definition and discussion of elements** *Video Vignettes** **Subtle QPQ** —Debrief questions with props/shoes *Role Plays** *Case Studies** *The Policy**	Low to Medium Risk Group Activities	35 minutes
Hostile Work Environment **Teaching Points** • Definition of Hostile Work Environment • Other Protected Categories • Whom Does the Policy Cover? • Welcome or Unwelcome Conduct—Who Decides?	*10 Danger Zones** *Role Plays** *Video Vignettes** *Case Studies** *The Policy** *Props**	Low to Medium Risk Group Activities	40 minutes

Content	Suggested Tools	Risk Activity Level	Estimated Time
• Severe or Pervasive • Not All Unwelcome Conduct Is Harassment • Off-Premises Conduct • "Bystander" Harassment • Gender-Based Harassment • Conduct That May Create a Hostile Work Environment			
The Reporting Process **Teaching Points** • The Organization Complaint Process • Confidentiality • Obligations of Managers and Supervisors • On Notice, Prompt Effective Action • Consequences	*The Policy Scavenger Hunt *Video Vignettes Debriefing *Role Plays —Responding to complaints *Case Studies *Workscripts *Props *Wallet cards	Low to Medium Risk	20 minutes
Special Obligations of Managers and Supervisors • Quid Pro Quo • Reporting Obligations • Taking in Complaints • Walking and Talking the Policy	*The Policy *Workscripts *Role Plays *Props *Case Studies	Low to Medium Risk	20 minutes
*Documentation • Attendance Sheet • Participation Form • Policy Acknowledgement Form	Documents	Low Risk	5 minutes
*Debrief Pre-Test	10 True/False Questions with communication cards	Low Risk	5 minutes
*Conclusion	*Tent Name Card Give away wallet card with reporting procedure on it	Low Risk	5 minutes

APPENDIX I

Trainer's Checklist for Manager's Course
Module 1

✔ TO DO BEFORE TRAINING

- ○ Fill in personalized paragraphs in Trainer's Manual
- ○ Memo/e-mail to participants regarding date, time, and place of training

✔ EQUIPMENT

- ○ TV/VCR or DVD Player (extension cord, if necessary)
- ○ Videotape or DVD for Blue-Collar Supervisors and Managers
- ○ Videotape or DVD for White-Collar Supervisors and Managers
- ○ Flip Chart and Easel/or Static Image Paper.
- ○ Cassette Player
- ○ Pens and Colored Highlighters
- ○ Heavy, Dark-Colored Markers for Writing on Flip Chart

✔ HANDOUTS

- ○ Attendance Form (one per session)
- ○ Participation Forms (one per participant)
- ○ Acknowledgement Forms (one per participant)
- ○ Our Anti-Harassment Policy (one per participant)
- ○ Our Other Relevant Policies (one per participant)
- ○ House Rules Tent Name Card (one per participant)
- ○ Manager Workbook (one per participant) (optional)
- ○ Evaluations (optional)

☑ PROPS (OPTIONAL)

- ○ 2 Shoes (Male and Female)
- ○ Music Cassette (instrumental)
- ○ Play Money (dollar bills)
- ○ Noisemakers
- ○ True/False Interactive Cards
- ○ Okay/Not Okay Interactive Cards
- ○ Money
- ○ Dice

☑ REFRESHMENTS

- ○ Soft Drinks, Coffee, and Snacks (optional)

APPENDIX J

Notable Verdicts and Settlements

Sexual Harassment and Sex Discrimination

A former Sprint Communications employee who claimed that he was sexually harassed by a male supervisor and suffered retaliation when he complained was awarded $1.2 million by a federal jury. *Thorne v. Sprint Communications,* Daily Lab. Rep. No. 65 (Apr. 4, 2002), A-3.

A federal judge in Washington gave final approval to a $31-million settlement of a nationwide sex discrimination suit against American Express Financial Advisors, Inc. An estimated class of up to 4,000 women could be affected by the decree, which would resolve allegations that female professional employees were paid less and given fewer job opportunities than their male counterparts. *Kosen v. American Express Financial Advisors, Inc.,* Daily Lab. Rep. No. 117 (Jun. 18, 2002), A-10.

A former Philadelphia firefighter who allegedly was retaliated against after complaining of sexual harassment by subordinates who believed he was gay has been awarded more than $1.2 million by a federal jury. *Bianchi v. City of Philadelphia,* Daily Labor Rep. No. 39 (Feb. 27, 2002), A-4.

A California jury awarded $30.6 million to 6 female employees who claimed they were subject to verbal and physical abuse by a former Ralph's Grocery Company store director. *Goldberg v. Ralph's Grocery Company,* Daily Lab. Rep. No. 69 (Apr. 10, 2002), A-10.

Plaintiffs' attorneys and the EEOC have reached a tentative agreement to settle a class action sexual discrimination suit against Rent-a-Center that includes a cash settlement of $47 million and an agreement by the national firm to make sweeping changes in its employment policies. *Wilfong v. Rent-a-Center,* Daily Lab. Rep. No. 47 (Mar. 11, 2002), A-1.

A federal jury awarded $3 million to a female patrol officer who alleged her employer, the Forest Preserve of Cook County, tolerated a sexually hostile work environment and then retaliated against her when she complained.

Spina v. Forest Preserve of Cook County, Daily Lab. Rep. No. 243 (December 20, 2001), A-5.

On June 13, 2001, a federal judge in New York refused to throw out a jury's decision to award $1.25 million to a New York City police officer repeatedly harassed, criticized, scrutinized, and eventually ousted in retaliation for complaining about sexual harassment. The court also added an additional $110,000 in punitive damages against three police department officials, and attorneys' fees and costs. *Gonzalez v. Police Commissioner Bratton,* Daily Lab. Rep. No. 119 (June 21, 2001), A-1.

More than 300 current and former female custodial workers employed by the Architect of the Capital have obtained court approval of a $2.5-million settlement resulting from a pay equity lawsuit filed against the AOC by a group of African-American female custodians. *Harris v. Architect of The Capital,* Daily Lab. Rep. No. 220 (November 16, 2001), A-8.

A federal jury awarded $2.2 million to a former employee of Outback Steakhouse, who alleged the restaurant chain discriminated against her by paying a male counterpart a significantly larger salary and then firing her after she complained. *EEOC v. Outback Steakhouse Inc.,* Daily Lab. Rep. No. 183 (September 24, 2001), A-7.

A federal judge in Florida reinstated his 1998 ruling that a Miami Beach restaurant owes $154,205 plus interest to fourteen female food servers who were denied jobs because the restaurant preferred a European ambiance with all male servers. *EEOC v. Joe's Stone Crab, Inc.* Daily Lab. Rep. No. 61 (March 29, 2001), A-6.

A federal jury in Kansas City has awarded a former Consolidated Freightways Corporation employee $1.135 million in her 1995 sex discrimination lawsuit against the company. *Lawrence v. Consolidated Freightways Corporation,* Daily Lab. Rep. No. 74 (April 17, 2001), A-7.

The EEOC has reached a $220,000 settlement on a sex discrimination lawsuit alleging the employer used a discriminatory pre-employment strength test for applicants for meter reader and janitor positions. According to EEOC, the strength test had the effect of excluding a disproportionate number of females from entry-level meter reader and janitor jobs. EEOC contended the test was unrelated to the entry-level jobs and only relevant to measure ability to perform some higher-level jobs that required certain strengths. *EEOC v. Consumers Energy Company,* Daily Lab. Rep. No. 74 (April 17, 2001), A-6.

The Second Circuit ruled that a criminal justice professor at Marist College was entitled to nearly $118,000 in back pay, liquidated damages, attorneys' fees, and costs for the college's violation of the Equal Pay Act. *Lavin-McEleney v. Marist College,* Daily Lab. Rep. No. 32 (February 15, 2001), A-6.

The Wisconsin Glass Container Company will pay a total of $180,000 in compensatory damages and attorneys' fees to five women who were allegedly sexually harassed on the job by another female worker under a consent decree with the EEOC. The EEOC claimed that the women were physically intimidated by their co-worker for several years. *EEOC v. Graf,* Daily Lab. Rep. No. 35 (February 21, 2001), A-2.

Race and National Origin Discrimination

The EEOC and a group of private plaintiffs reached a $1.8-million consent decree with an Illinois manufacturer, settling racial harassment suits brought on behalf of 32 former and current African-American workers. *EEOC v. Scientific Colors, Inc., d/b/a Apollo Colors,* Daily Lab. Rep. No. 60 (Mar. 28, 2002), A-1.

Approximately 120 African-American current and former employees of NASA who allegedly were denied promotions because of their race will receive more than $2 million in back pay under a class action settlement agreement, which recently received preliminary approval from the EEOC. *Flournoy v. O'Keefe,* Daily Lab. Rep. No. 88 (May 7, 2002), A-7.

A cable company will pay $1.5 million to six of its employees who allegedly were subjected to a racially hostile work environment, under a consent decree the company arranged with the EEOC. *EEOC v. Adelphia Cable Partners,* Daily Lab. Rep. No. 87 (May 6, 2002), A-8.

A state agency responsible for regulating banks in Illinois faces a $1-million liability, after a federal judge approved a verdict finding that it had discriminated against two African-American employees. *Robinson v. Illinois Office of Banks and Real Estate,* Daily Lab. Rep. No. 85 (May 2, 2002), A-3.

A group of 139 black former employees of a Sara Lee Foods meat processing plant in Philadelphia will be paid $3.5 million under a private settlement of a race discrimination lawsuit filed three years ago. Daily Lab. Rep. No. 103 (May 29, 2002), A-2.

Ford Motor Company has agreed to pay $300,000 to a class of 23 African-American employees at one of its Detroit area plants to settle racial harassment charges filed by the EEOC. In its lawsuit, EEOC charged that a white former co-worker harassed plaintiff and 22 other similarly situated black employees. According to the commission, the employee hung Hangman nooses on his forklift, mocked his co-workers' speech and walking styles, and used insulting language aimed at African-Americans. Several times the worker said, "This is a good day for a lynching." The worker was subsequently disciplined and has retired from Ford. *EEOC v. Ford Motor Company,* Daily Lab. Rep. No. 245 (December 24, 2001), A-1.

A federal judge approved a $5.6-million settlement of a class action race discrimination lawsuit against Dillard's, Inc. Participants in the suit alleged that

African-American employees at Dillard's stores in the Kansas City region faced a hostile work environment and suffered discrimination in hiring, pay, promotion, and terms and conditions of employment. *Wooten v. Dillard's*, Daily Lab. Rep. No. 240 (December 17, 2001), A-9.

A federal judge has given final approval to a record $192.5-million settlement in a class action race discrimination lawsuit against Coca-Cola. The settlement also requires Coca-Cola to revamp its human resources policies and submit to outside oversight of its employment practices. As part of the settlement, Coca-Cola will allow close scrutiny of its human resources procedures for a four-year period. *Abdallah v. Coca-Cola Company*, Daily Lab. Rep. No. 127 (July 3, 2001), A-5.

On July 2, 2001, Beverly Enterprises Inc., one of the largest operators of nursing homes in the country, will pay $1.2 million to settle racial harassment allegations by the EEOC. The settlement stems from a lawsuit alleging that the administrator of the nursing center pursued a campaign of racial harassment in which she allegedly coded applications with smiling faces for white applicants and frowning faces for black applicants, frequently used racial slurs, and ordered excessive discipline of black employees. *EEOC v. Beverly Enterprises*, Daily Lab. Rep. No. 128 (July 5, 2001), A-1.

A Tacoma, Washington, car dealership will establish a fund of $750,000 to settle claims arising under a lawsuit that alleged the company subjected a class of African-American employees to persistent racial harassment. EEOC had filed suit against the car dealership a year prior to the settlement after employees with the company charged that management regularly used racially offensive language and made management decisions based on the race of employees and customers. Racially derogatory cartoons, jokes, and slurs occurred over a long period of time. In an earlier settlement in the case, the dealership agreed to pay $1.15 million to settle claims of individuals who intervened in the EEOC lawsuit. *EEOC v. Robert Larsons Chrysler-Plymouth of Tacoma, Inc.*, Daily Lab. Rep. No. 28 (Feb. 9, 2001), A-6.

On March 29, 2001, a federal judge in Peoria gave preliminary approval to an agreement settling a class action suit alleging racial discrimination and harassment by Mitsubishi Motor Manufacturing of America. The settlement provides $1.4 million to 10 named plaintiffs in the case and creates a structure to which damages will be provided to as many as 250 additional class members. The settlement also provides $1.8 million to plaintiffs' counsel for fees and litigation expenses. Daily Lab. Rep. No. 63 (April 2, 2001), A-10.

Georgia Pacific Corporation reached a $200,000 settlement of a racial harassment suit by the EEOC, in a consent decree approved by a federal judge in North Carolina. Under the terms of the agreement, the company will pay $200,000 to a class of African-American employees who work in the fabrication shop at the company's North Carolina facility. The company also agreed to training on anti-discrimination laws for employees and management at the facility. Daily Lab. Rep. No. 65 (April 4, 2001), A-11.

It was announced on March 11, 2001, that the Alabama Department of Transportation reached a $65.1-million settlement in its fifteen-year-old race discrimination case. The settlement was in response to a class action alleging race discrimination in hiring, promotions, compensation, and other opportunities. The agreement will settle claims with about 2,400 employees and more than 100,000 applicants. According to the proposed settlement, $40 million would be paid for promotion and compensation claims, up to $15 million would be paid to settle hiring claims, and at least $4.6 million would be paid to settle compensatory damage aspects of the plaintiffs' contempt claims. *Reynolds v. Alabama Dept. of Transportation*, Daily Lab. Rep. No. 54 (March 20, 2001), A-3.

A federal jury in Oregon awarded $1.4 million in compensatory and punitive damages to a trainee in a police department who alleged race discrimination. Plaintiff claimed that he suffered racial profiling and that his performance evaluation was falsely scored. *Bell v. Clackamas County*, Daily Lab. Rep. No. 29 (Feb. 12, 2001), A-6.

A federal jury awarded nine white Connecticut firefighters $3.2 million after finding that they had been the victims of reverse discrimination. *Petersen v. Hartford*, Daily Lab. Rep. No. 82 (Apr. 29, 2002), A-8.

A jury ordered Fulton County, Georgia, to pay $25 million to eight white librarians it believed were discriminated against because of their race when they were transferred from a main library to branch offices. *Bogle v. McClure*, Daily Lab. Rep. No. 14 (Jan. 22, 2002), A-1.

Salomon Smith Barney, Inc., will pay $635,000 and offer salary increases and enhanced promotional activities to a class of computer employees claiming employment discrimination on the basis of race and national origin, according to the terms of a settlement with the EEOC. The firm was also required to continue workforce education on employment discrimination—with all managers and supervisors to receive equal employment opportunity training at least annually—and to submit a written report to EEOC within ten days of the training, describing it and documenting attendance. *EEOC v. Salomon Smith Barney, Inc.*, Daily Lab. Rep. No. 136 (July 17, 2001), A-11.

A U.S. district judge approved a jury's award of $3 million in damages and back pay to a Chinese immigrant for discrimination based on race and national origin by his former employer, a Seattle seafood firm. *Zhang v. American Gem Seafoods*, Daily Lab. Rep. No. 194 (October 10, 2001), A-4.

Prudential Insurance Company has agreed to pay $300,000 to four Haitian employees who were told by supervisors not to speak Creole, their native language, according to a consent decree filed in federal court. *EEOC v. Prudential Ins. Co. of America*, Daily Lab. Rep. No. 165 (August 27, 2001), A-2.

Age Discrimination

On March 14, 2002, a Michigan judge granted final approval to a $10.5-million settlement of claims alleging that Ford Motor Company's performance evaluation system discriminated against older managers. *Siegel v. Ford Motor Co.*, Daily Lab. Rep. No. 51 (Mar. 15, 2002), A-1.

Kraft Foods will pay $270,000 to a former salesman at Nabisco, Inc., who two years ago was fired and replaced with a younger worker during a reorganization, under a consent decree reached with the EEOC. *EEOC v. Kraft Foods North America, Inc.*, Daily Lab. Rep. No. 80 (Apr. 25, 2002), A-2.

A New Jersey jury awarded $1 million in damages for back pay and emotional distress to a former bank branch manager who claimed that her age was the sole reason she was fired in 1997. *O'Shea v. Summit Bancorp*, Daily Lab. Rep. No. 81 (Apr. 26, 2002), A-6.

A New Jersey Superior Court jury awarded $3.9 million in compensatory and punitive damages to a paper company executive who sued his former employer for age bias and retaliation after he was let go in a company reorganization. *Zacharias v. Whaman*, Daily Lab. Rep. No. 111 (June 10, 2002), A-4.

A former sales manager for Abbott Laboratories was awarded $25 million in punitive damages by a Columbus jury for age discrimination. The judge said the jurors were concerned about the amount of the award, but believed it was appropriate based on the company's resources. Abbott had $16.3 billion in sales and $2.9 billion in net earnings in 2001. *Jelinek v. Abbott Laboratories*, Daily Lab. Rep. No. 86 (May 3, 2002), A-7.

A federal district court found a Denver TV station guilty of age discrimination and ordered the station to pay $562,152 to a fired news reporter. According to court testimony, the new management team decided that the station, which had the lowest ratings of any Denver network news affiliate, needed changes to attract a younger audience. *Minshall v. McGraw-Hill Broadcasting Co.*, Daily Lab. Rep. No. 174 (September 10, 2001), A-15.

Disability Discrimination

The EEOC won a $185,000 verdict in an ADA suit on behalf of a man who died of stomach cancer shortly after being terminated by a Chicago-based security services company. While the jury granted plaintiff only $961 in back pay and $187 in compensatory damages, they awarded him $184,000 in punitive damages. *EEOC v. Mid-Continent Security Agency, Inc.*, Daily Lab. Rep. No. 68 (April 9, 2001), A-1.

The Fifth Circuit upheld an award of $590,000 to a former General Electric Company machinist on his ADA claim. The court noted that seeking SSDI benefits did not conflict with his claim that he was qualified to work under the ADA. *Giles v. General Electric Company,* Daily Lab. Rep. No. 44 (March 6, 2001), A-1.

A federal jury in Chicago awarded $1.45 million to a woman who claimed that her employer discriminated against her because of disabilities stemming from lupus and depression. The jury concluded that the defendant-employer subjected plaintiff to a hostile work environment, failed to accommodate her disabilities, and constructively discharged her. *Swiech v. Gottlieb Memorial Hospital,* Daily Lab. Rep. No. 44 (March 6, 2001), A-6.

Others

Two Muslim contract employees of Motorola will share a $60,000 consent decree settling a suit filed by the EEOC that accused Motorola of religious discrimination. *EEOC v. Motorola, Inc.,* Daily Lab. Rep. No. 9 (Jan. 14, 2002), A-14.

A California court affirmed a $2.66-million verdict to a former Oracle vice president who claimed both retaliatory discharge for complaining about pregnancy discrimination and wrongful discharge for complaining about other workers' alleged inappropriate use of an affiliate's computer software. *Baratta v. Oracle Corp.,* Daily Lab. Rep. No. 31 (Feb. 14, 2002), A-2.

A male basketball and soccer coach at all-women Smith College has been ordered $1.65 million by a Massachusetts jury that found sex and age bias played a role in his termination. *Babyak v. Smith College,* Daily Lab. Rep. No. 245 (December 24, 2001), A-3.

A federal jury in Washington, D.C., awarded $1.7 million in damages to a transit authority manager who claimed he was denied a promotion in retaliation for testifying in favor of a plaintiff claiming discrimination on the part of the agency. *Mackel v. Washington Metro Area Transit Authority,* Daily Lab. Rep. No. 213 (November 6, 2001), A-2.

General Motors Corporation agreed to pay 16 employees $1.25 million to settle sexual and racial harassment charges, filed by the EEOC as part of a consent decree. *EEOC v. General Motors Corp. Inc.,* Daily Lab Rep. No. 186 (September 27, 2001), A-1.

Eagle Global Logistics agreed to pay $9 million to settle claims of discrimination against minorities and women under a consent decree with the EEOC. The EEOC alleged that the company failed and/or refused to promote African-Americans, Hispanics, and female employees into managerial and sales positions and paid these workers less than Caucasian and/or male employees for performing similar or comparable work. *Dube v. Eagle Global Logistics,* Daily Lab. Rep. No. 190 (October 3, 2001), A-13.

Approximately 400 job applicants and 64 current and former employees of Ingersoll Milling Machine Company will share in a $1.8-million settlement, which resolves race and sex discrimination allegations brought by the EEOC. The consent decree also requires the company to conduct equal employment training and recruit a diverse field of job applicants. *EEOC v. Ingersoll Milling Machine Co.*, Daily Lab. Rep. No. 136 (July 17, 2001), A-7.

The First Circuit Court of Appeals on September 4, 2001, affirmed a $730,000 jury verdict in favor of a credit union financial planning manager, which included $400,000 in punitive damages and $200,000 in compensatory damages on her retaliation claim. *Zimmerman v. Direct Federal Credit Union,* Daily Lab. Rep. No. 172 (Sept. 6, 2001), A-5.

On July 10, 2001, Wal-Mart entered into a consent decree with the EEOC, requiring the retailer to pay a former manager $55,000 to resolve charges of age discrimination and retaliation. In addition, Wal-Mart agreed to retrain all salaried and hourly supervisory management personnel employed in a five-store district in the southwest suburbs of Chicago. The training will be conducted by Wal-Mart personnel and will focus on the ADEA and Wal-Mart's policies barring discrimination and retaliation. *EEOC v. Wal-Mart Stores Inc.,* Daily Lab. Rep. No. 132 (July 11, 2001), A-7.

A jury in Maryland awarded more than $11,000,000 to a trucking manager who sued Safeway Inc. for defamation based on the company's handling of a co-worker's complaint of sexual harassment. In addition, plaintiff's wife was awarded $500 for damage to their marriage. According to plaintiff's attorney, the size of the verdict was based upon the jury being quite upset with Safeway's negligent investigation of the harassment complaint. *Talley v. Safeway Inc.,* Daily Lab. Rep. No. 22 (Feb. 1, 2001), A-6.

Approximately 2,200 black male current and former Social Security Administration employees and their attorneys will receive $7.75 million to settle their race and sex discrimination class action. *Burden v. Barnhart,* Daily Lab. Rep. No. 11 (Jan. 16, 2002), A-1.

Optical Cable Corp. reached a $1-million consent decree with the EEOC, settling charges of race and sex discrimination. *EEOC v. Optical Cable Corp.,* Daily Lab. Rep. No. 37 (Feb. 25, 2002), A-1.

APPENDIX K

Attendance Sheet

(page _____ of _____)

Employer Name: _____

Date: _____   **Trainer(s):** _____

Location: _____

Directions for Participants: **Signature Below**
Please <u>print</u> your full name below:

Print name below and then sign next to it.

_____   _____

_____   _____

_____   _____

_____   _____

_____   _____

_____   _____

_____   _____

_____   _____

_____   _____

_____   _____

_____   _____

_____   _____

APPENDIX L

Evaluating the Training

**Company
Name:** _____

Date: _____

Trainer: _____

1. On a scale of 1 to 7 (with 7 being high), how would you rate the overall quality of this course?

   _____

2. What did you find **most** helpful about this course?

   _____

   _____

3. What did you find **least** helpful about this course?

   _____

   _____

4. Please comment on the course leader's effectiveness in terms of style, facilitation, responsiveness to participants, and knowledge of the subject.

   _____

   _____

5. What changes, if any, would you suggest to today's course?

_____

_____

6. Comments: _____

_____

_____

APPENDIX M

Participation Form

Directions:

Please read and fill out this page, then give it to the leader(s) before the end of the training program.

I, _____ (Print Name), participated in the company's

harassment training program on _____ (date).

X _____
　　　　　　　　　Signature of Participant

X _____
　　　　　　(Print Name: First, Middle Initial, Last)

_____
Employee No. (if applicable)

_____
Employee Title (if applicable)

APPENDIX N

Acknowledgment Form
of Anti-Harassment Policy

I acknowledge that I have been given a copy of the Anti-Harassment Policy. I have read it, I understand its terms and procedures, and I have no questions about it. Furthermore, I agree to abide by it, and I understand that if Client determines my conduct so warrants it, I may be subject to discipline pursuant to the Policy including the immediate termination of my employment.

_____

Employee Name (Please Print)

_____

Employee Signature

_____

Date

Index

Books from the American Bar Association
Section of State and Local Government Law

ABCs of Arbitrage, 2002 Edition
Tax Rules for Investment of Bond Proceeds by Municipalities
Frederic L. Ballard, Jr.
Concentrates on tax-exempt bonds issued by a municipality and the market for taxable bonds issued by a corporation or by the federal government. *2002, 7 x 10, 385 pages, paper*

PC: 5330078   *SLGL member price: $119.95*   *Regular price: $134.95*

Census 2000: Considerations and Strategies for State and Local Government
Benjamin E. Griffith, Editor
Highlights the benefits and advantages of the Census Bureau's strategic planning and focuses on the federal, state and local government partnerships that will distinguish Census 2000 from prior census programs. *2000, 6 x 9, 208 pages, paper*

PC: 5330072   *SLGL member price: $59.95*   *Regular price: $69.95*

Court-Awarded Attorneys' Fees: Examining Issues of Delay, Payment and Risk
By Russell E. Lovell, II
Provides an in-depth examination of the history, process and structure of these fees.
1999, 6 x 9, 288 pages, paper

PC: 5330070   *SLGL member price: $69.95*   *Regular price: $79.95*

Ethical Standards in the Public Sector
A Guide for Government Lawyers, Clients and Public Officials
Patricia E. Salkin, Editor
A compilation of essays, articles, and research intended to help government lawyers, their clients, and other public officials focus on some of the ethical considerations that arise in the practice of law in the public sector. *1999, 7 x 10, 352 pages, paper*

PC: 5330067   *SLGL member price: $74.95*   *Regular price: $84.95*

Freedom of Speech in the Public Workplace
A Legal and Practical Guide to Issues Affecting Public Employment
Marcy S. Edwards, Jill Leka, James Baird, Stefanie Lee Black
Includes discussions about how the First Amendment applies to the speech of public employees, how race, sex, sexual preference, and religion are affected by the First Amendment and what issues are relevant when public functions are privatized. *1997, 7 x 10, 200 pages, paper*

PC: 5330063   *SLGL member price: $64.95*   *Regular price: 74.95*

From Sprawl to Smart Growth
Successful Legal, Planning, and Environmental Systems
Robert H. Freilich, Editor
Discusses how states and local governments can control sprawl, maintain urban areas, enlarge their quality of life through new urban and mixed use developments, and increase the economic development base through transportation corridors with joint public-private development while ensuring a sustainable environmental and agricultural way of life.
2000, 7 x 10, 368 pages, paper

PC: 5330068   *SLGL member price: $79.95*   *Regular price: $89.95*

Hot Topics in Land Use Law: From the Comprehensive Plan to Del Monte Dunes
Patricia E. Salkin and Robert H. Freilich, Editors
This book combines an array of land use articles from *The Urban Lawyer* with specially commissioned essays. Covers a broad range of issues that are changing the way land use law and zoning are practiced and interpreted in the courts. *2000, 6 x 9, 223 pages, paper*
PC: 5330069   *SLGL member price: $59.95*   *Regular price: $69.95*

How to Litigate a Land Use Case: Strategies and Trial Tactics
Larry J. Smith, Editor
A step-by-step guide from the earliest days of a land use controversy to the delivery of the oral argument in an appellate court. *2000, 6 x 9, 426 pages, paper*
PC: 5330073   *SLGL member price: $64.95*   *Regular price: $74.95*

Sexual Harassment in the Public Workplace
Benjamin E. Griffith, Editor
Provides an in-depth analysis of the most current caselaw, and information on how to try and defend sexual harassment cases. *2001, 6 x 9, 312 pages, paper*
PC: 5330074   *SLGL member price: $75.00*   *Regular price: $90.00*

Sword & Shield Revisited, Second Edition: A Practical Approach to Section 1983
Mary Massaron Ross, Editor
Provides an understanding of the framework of Section 1983 and takes you beyond the basic issues to the comprehensive information to assist you in your practice. *1997, 695 pages, 6 x 9, paper*
PC: 5330064   *SLGL member price: $75.00*   *Regular price: $85.00*

Taking Sides on Takings Issues: Public and Private Perspectives
Thomas E. Roberts
Compiles and contrasts the public and private perspectives on the most controversial issues in takings law. *2002, 6 x 9, 600 pages, paper*
PC: 5330077   *SLGL member price: $75.95*   *Regular price: $90.00*

Taking Sides on Takings Issues: The Impact of Tahoe-Sierra
Thomas E. Roberts
Offers a comprehensive analysis of the blockbuster takings case, *Tahoe-Sierra Preservation Council v. Tahoe Regional Planning Agency. 2003, 6 x 9, 109 pages, paper*
PC: 5330079   *SLGL member price: $29.95*   *Regular price: $34.95*

Trends in Land Use Law from A to Z: Adult Uses to Zoning
Patricia E. Salkin, Editor
Includes information on *Palazzolo v. State of Rhode Island*, the short decision handed down by the U.S. Supreme Court in the *Olech* case, and the issues that arise when land use law meets the First Amendment. *2001, 6 x 9, 504 pages, paper*
PC: 5330075   *SLGL member price: $75.00*   *Regular price: $110.00*

For more information or to order these books, please call (800) 285-2221 or visit our web site at www.ababooks.org